D0579485

Great Caves of the World

TONY WALTHAM

FIREFLY BOOKS

3 1232 00848 2640

Published by Firefly Books Ltd. 2008

First published by the Natural History Museum, London

First printing

Publisher Cataloging-in-Publication Data (U.S.)

Waltham, Tony.
   Great caves of the world / Tony Waltham.
[112] p. : col. photos. ; cm.
Summary: Thirty of the world's most spectacular, challenging
and diverse caves. Each entry includes details of the cave's
inhabitants and environment, how and when it was discovered, how
it was formed, and, where accessible, tips on how to travel to the cave.
ISBN-13: 978-1-55407-413-6
ISBN-10: 1-55407-413-4
1. Caves. I. Title.
910.02144 dc22   GB601.3.W358 2008

Library and Archives Canada Cataloguing in Publication

Waltham, Tony
   Great caves of the world / Tony Waltham.
ISBN-13: 978-1-55407-413-6
ISBN-10: 1-55407-413-4
1. Caves. I. Title.
GB601.3.W34 2008   910'.02144   C2008-901276-3

Published in the United States by
Firefly Books (U.S.) Inc.
P.O. Box 1338, Ellicott Station
Buffalo, New York 14205

Published in Canada by
Firefly Books Ltd.
66 Leek Crescent
Richmond Hill, Ontario L4B 1H1

Written by Tony Waltham
Designed by Studio Gossett
Reproduction by Saxon Photolitho, UK

Cover photo credits
Front cover: Gruta do Janelão, Brazil © Tony Waltham Geophotos
Back cover: Skocjanske Jame, Slovenia © Park Skocjan Caves, Slovenia.
   Photo: Borut Peric

Printed in China

Introduction

Throughout the world, caves lure cavers and geologists alike to discover how far they extend underground and just how they were formed. The caves described in this book each have some special feature, whether it is their sheer size, their memorable beauty, the mystery of their origins or even the origins of life.

Below: **Phreatic cave**

The succession of rounded domes and arches along the roof of the Niah Great Cave, in Sarawak, show that it was formed as a phreatic cave long ago when it was completely filled with water.

Most caves are formed in limestone, beneath streamless landscapes known as karst. Rainwater slowly dissolves the limestone as it flows down through the bedrock joints. It can create a cave passage wherever it finds a drainage route through a limestone hill, from a sinkhole or a valley floor through to an outlet at a spring. Water enters a cave through narrow fissures or huge open shafts, and leaves through a resurgence, that is either an open passage or a flooded tunnel beneath a lake surface.

The shapes and profiles of cave passages, and entire cave systems, depend partly on the pattern of joints and bedding planes that the underground drainage originally followed, but more significantly on how the passages were subsequently enlarged. Where a cave passage lies above its resurgence, a stream can freely drain through it. The tumbling and swirling water erodes the cave floor and cuts down into a narrow, twisting, canyon passage, interrupted by waterfall shafts that are enlarged by spray erosion. This is a vadose cave, because it lies above the water table. Below the water table, any

passage is totally flooded because its water backs up to its overflow point at the level of the resurgence. As a result, its roof is eroded just as much as its floor, creating a passage with a tubular shape that can extend uphill or downhill because water can flow in any direction under pressure. This is a phreatic cave, meaning that it is full of water - or was full of water when it was formed. Many vadose streamways end at dark sump pools, where the passages only continue underwater as phreatic tunnels.

Cave passages evolve. Landscapes undergo ceaseless erosion that results in surface lowering. Lower valley floors allow new resurgences to develop, and underground streams find their way down through the rock to create new cave passages at lower levels. High-level caves are then abandoned; whether they were once roaring canyon streamways or silent flooded tubes, they can be left high and dry. These abandoned caves are records of past events, because they show where streams once flowed and at what altitudes valley floors once stood. The caves themselves cannot be dated, but many of their sediments can be, mainly by measuring the ratios of radioactive isotopes in stalagmites and cave sediments. Most caves have developed within the last million or so years, when climates and environments fluctuated through the Ice Ages.

[Not just limestone and rainwater] Rainwater alone can

dissolve limestone, but it is far more effective after it has passed through a soil layer so that it contains additional carbon dioxide derived from plant decomposition. Most limestone dissolution depends on carbon dioxide, so more plants mean bigger caves - which is why so many of the world's giant caves lie beneath the forests of the wet tropics. Some very large caves in desert or polar regions are relics from contrasting climates in the past. But others have a completely different origin, some were formed by sulphuric acids rising from deep

Above: **Vadose cave**
This spectacular vadose shaft was carved into its cylindrical shape largely by spray when it was occupied by a waterfall that was an unbroken 500 m (1640 ft) deep, inside the Miao Keng cave in the karst of Chongqing, China.

basins of sedimentary rock. There is still much debate over how some of the world's caves were formed in limestone.

Caves can also form in other rocks, notably gypsum and salt, which are both soluble in water. There are some very extensive gypsum caves scattered around the world, though salt caves are rare because the rock is so soluble that it only survives in dry desert terrains. Caves are also found in lava, where they formed not by any form of erosion but by hot molten rock flowing out of the cores of lava flows.

[Underground decorations]
The same seeping water that creates caves by dissolving rock may also fill them up by deposition of calcite - the mineral that comprises both limestone and stalagmites and stalactites. Both processes depend on the levels of carbon dioxide in the rainwater, in the soil and in the cave air. Under the right conditions, water that drips from the cave roof deposits calcite to create the host of decorations that adorn so many cave passages. Stalactites hang from the roof, stalagmites grow up from the floor, and columns are created where the two join. All can be beautifully white, or tinted yellow (mainly by iron oxides). A special type of stalactite is the straw, which has the proportions of a drinking straw and is formed where calcite is deposited just round the rim of water drops that repeatedly hang from its tip. More bizarre is the tiny helictite, which grows in totally random shapes - upwards, outwards and sideways - guided by crystal growth within its structure, though there is still some mystery over the causes of their shapes.

A cave stream erodes its floor, but films of water saturated in lime can deposit calcite flowstone on a cave floor. This commonly spreads out from stalagmites and can build up in multiple layers. Rimstone pools, also known as gour pools, lie behind their own calcite dams that have built up by calcite deposition from their thin films of overflow water. Some pools have rimstone dams that grow inwards as shelfstone deposited at the water level, while others have thin, floating rafts of calcite held up by surface tension. Within the pools, calcite may be deposited as rounded masses or as sharp crystals known as dog-tooth spar. As a further variation, drips landing in a shallow pool can disturb the water so that grains on its floor are rolled around, and deposition of the calcite on all sides creates perfectly rounded cave pearls.

Not all decorations are made of calcite. Aragonite has the same chemistry but a different molecular structure, and grows as either sharp and delicate crystals or as tiny branching trees, generally where water evaporates off a calcite surface. Gypsum decorations are formed where sulphate-rich waters seep into limestone caves; their crystals grow from the base, like animal hair, so gypsum flowers form where clusters of crystals curve and splay outwards from a cave wall. Not quite a mineral, ice grows in caves that are cold enough, where both giant frost crystals and dripwater icicles can produce some of the most fantastic cave decorations.

[Once formed, not forgotten] Smoothly sculpted and rounded

rock walls are the signatures of underground streams that carved caves out of solid rock. But dissolving away the rock is only the start of a cave's history. Roof collapse modifies the profiles of many caves where blocks of rock fall away from unsupported ceilings. Unless removed by a stream, these piles of breakdown can completely block a passage. Collapse cannot form cave chambers on its own, but it modifies their profiles and may cause adjacent cave passages to coalesce into larger voids. Breakdown debris is joined by sand, gravel and mud that are commonly washed into caves. All of these can accumulate in passages, often combining with calcite deposits to partially or totally block them. Many entrances are choked by stream debris, hillside rubble and even glacial debris, so that cavers have to dig a way through before starting their underground explorations.

As an environment for underground life, most caves suffer from a shortage of available food. But a large cave in a tropical climate can house huge colonies of bats that fly out nightly to feed on insects. Bat guano, the droppings beneath their roosts, is the food source for huge assemblages of bugs, beetles, spiders and many more cave-animals. The guano may also be mined to produce fertiliser, but man has long had other uses for caves. Stone Age man lived in the entrance zones, and some ventured further in to paint animal figures on the walls. Some caves have long been holy sites, and others have later been developed as tourist sites. Through all these stages, caves have developed and matured into innumerable variations, making them a fascinating part of the natural world.

INTRODUCTION

Below: **Dripstone straw**
A vertical straw stalactite is little wider than a drinking straw, but supports a series of tiny helictites growing on its right side so that they all point downwind; these are in the small Polje Cave on Qeshm Island, Iran, and are made of salt even though they mimic the forms usually made of calcite.

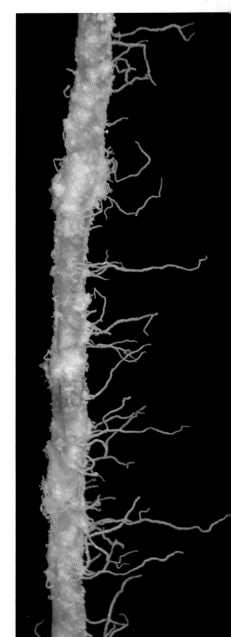

Sof Omar Cave ▮ Ethiopia

[MAZE CAVE WITH A RIVER]

A LAVA PLATEAU IN THE EASTERN HIGHLANDS of Ethiopia might seem the last place to find a large cave. But the lava is only a thin veneer over an extensive bed of limestone, and within this, far from any other known cave, lies one of the world's great underground rivers. A splendid box-shaped canyon cuts through the plateau, with rims of lava dropping steeply to canyons cut down into the limestone. The Webi Gestro River drains off the Bale Mountains to flow along the canyon floor, except at one place where a great sweeping meander is always dry. The river takes a short-cut underground, draining through the limestone neck of the meander in the magnificent cave of Sof Omar, named after the little market village on the plateau almost above it.

To refer to the 'river sink' is rather an inadequate description for this grand cave entrance, where the entire river sweeps into darkness in a magnificent tunnel in clean, pale limestone. The rock has been carved into pillars and alcoves by eons of dissolution (see p.5), and is marred by neither mud nor stalactites. The site has been known to generations of the local farmers as a valuable source of water, and there are usually camels, cows and goats drinking or cooling off in the large pool immediately outside the cave mouth. A cluster of small mud-and-thatch houses on the adjacent bank adds to a classic scene of rural Africa, but to most of the local people the cave is nothing more than the end of the river.

It was visiting groups of Europeans who, together with a few of the more adventurous local lads, bothered to go further into the cave, and eventually mapped its 15 km (9 miles) of passages. They found a splendid river gallery, which keeps its size of 10-20 m (30-60 ft) high and wide for the entire two kilometres through to its resurgence (see p.4). With only a gentle gradient, the cave river runs over a floor of black lava cobbles broken by low cascades over a few ledges of bedrock limestone. Pools and lakes are long and shallow except for a few that catch the unwary caver with surprisingly deep water. The high roof opens up into a few rounded domes, but there is little rock collapse along the way.

Sof Omar Cave exists because the river found an easy route following beds of limestone that dip just gently though the neck of the valley's meander. A river always favours the steeper gradient of the shortest route between two points. Here, it found networks of fractures that offered a route through one thin zone of beds; once established as a flow route, dissolution and erosion of the pure limestone soon etched out a grand cave. In the annual rainy season, the river hurtles through the cave 6 m (20 ft) deep, scouring clean the rock walls.

[**The great maze**] Where the Webi Gestro has found its way through the limestone hill, it has not entirely confined itself to a single route. Two sets of intersecting joints in the limestone have offered a vast number of alternative routes, and many of these have been etched out to form a complex network of passages that now form the great underground maze of Sof Omar. Mazes are well-known features of caves that regularly flood, when overflow water finds its way through as many routes as possible. Sof Omar's maze has developed through hundreds of thousands of years, so many of its passages are now permanently dry - left as abandoned tunnels off the side of the active river gallery. The maze is a network of amazing complexity, which can baffle the casual visitor. Long tunnels and innumerable crossroads break out into large chambers and in parts leave only slender pillars of bedrock between adjacent passages.

Sof Omar is far from anywhere, so few visitors pass by, but those that venture through the cave are in for a real treat. Young men of the village are quick to offer their services as guides, who are more than useful in the cave's complexities. They weave their visitors on a clever and varied route through parts of the maze and sections of the river passage. And the spectacular climax is at the end of the underground journey, where they emerge from the maze though the walls of the downstream canyon. The multiple passages on the many joints allow daylight to pour into the cave, and the play of light and shadow is totally beautiful in a long transition zone from dark cave to African sunshine. A visit to Sof Omar is truly inspiring.

Opposite: **Maze exit**
Daylight streams into the Sof Omar Cave from the canyon above its resurgence. All the cave passages follow just a few beds within the limestone sequence; they also follow major joints that intersect to create the network grid of the main maze.

Sterkfontein Cave ▮ South Africa

[THE CAVE OF MAN'S EARLIEST ANCESTORS]

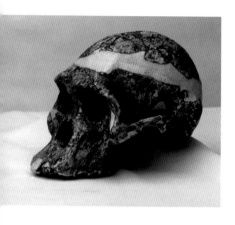

Above: **Mrs Ples**

The almost complete skull of the fossil hominid known as Mrs Ples, which was blasted out of the hard cave breccia of Sterkfontein in 1947.

Opposite: **Rift chambers**

A small lake in the lower chamber of Sterkfontein Cave hides totally flooded rifts that extend through to further dry passages.

A LITTLE HILL, IN THE GRASSLANDS west of Johannesburg, is scored by a few crags that show it is made of bedrock. Between the crags, archaeologists probe soil-filled fissures, and a path leads up to the entrance to Sterkfontein Cave. Along with a handful of other caves in the same area, now dubbed the 'Cradle of Humankind', it has yielded about a third of the world's known fossils of early hominids - the true ancestors of humans.

The cave is a series of tall rift chambers enlarged from vertical joints, leaving between them fretted pillars of hard dolomitic limestone. It was formed when it was completely under water, but the water table has now declined, exposing the dry chambers, while the lower rifts still continue down into deep water. Some millions of years ago, the tops of the cave rifts were exposed by surface lowering, so that debris fell into the cave and slowly accumulated in great conical piles. The lower parts of these have since been re-cemented by natural calcite deposition to form a hard breccia rock, and this contains a wealth of fossils. These were bones dropped by carnivores that lived among the trees around the open cave shafts. One of the oldest fossils is the disjointed, but nearly complete, skeleton known as Little Foot. This is a very early hominid, closely related to the apes but one that did walk upright. Found at a slightly higher level, a well-preserved skull of *Australopithecus africanus* has a profile that places it between those of apes and humans. Named Mrs Ples after its earlier name of *Plesianthropus*, it was first thought to be a woman, but recent studies of the teeth imply that this is the skull of a young male.

Dating these fossils is difficult because the layering of the breccia is so complex. Though originally thought to be older, Little Foot's age has recently been confirmed as about 2.2 million years, and Mrs Ples appears to be just a little younger. Fossils this good are rarely preserved outside a cave, and both are part of an unrivalled sequence of hominid fossils that show the early evolution of mankind. They make Sterkfontein a very valuable site, where there is probably yet more to discover by further excavations.

Castleguard Cave ▌ Canada

Opposite: The Subway

The perfect phreatic tube known as The Subway; its floor is scored by a tiny vadose trench, formed by modern seepage water.

HIGH IN THE CANADIAN ROCKIES, and not far west of the Banff-Jasper Highway, the Columbia Icefield lies in a high-level basin, ringed by peaks that include Mt Columbia and Mt Castleguard. Long ago, this basin had no icefield; rainwater and snow melt sank into its bare limestone floor, and drained out through Castleguard Cave. The cave was a single long phreatic loop (see p.5); the only outlet for the water was a spring in the floor of the contemporary Castleguard Valley. Then, within the last million years or so, glaciers and interglacial rivers eroded the valley to its present level and cut away the downstream end of the cave, to leave an open entrance in the hillside.

Most of the water in the cave drained out through new springs in the new valley floor, and the once-flooded phreatic tubes were largely transformed into dry tunnels; the best of these is known today as The Subway. In the floors of many of the tubes, small vadose streams (see p.4) cut deep and narrow trenches that are now known as the Fissures.

Castleguard's water comes from summer melt from the icefield, so flow ceases in winter, which is therefore the only safe time to visit the cave. Inside, the rock is clean and dry, and frozen pools are domed by the pressure of expanding ice. Below a short drop, the next 300 m (1000 ft) of level passage are filled with ice nearly to the roof; the Ice Crawls require half an hour of squirming along, flat-out between ice and rock. The next pools are unfrozen - the passage is now far enough below the surface that geothermal heat keeps its temperature just above freezing. A shaft of 25 m (80 ft) drops into The Subway. This gun-barrel-like passage, 3 m (10 ft) in diameter, is a perfect example of a drained phreatic tube, and is the best evidence of the cave's phreatic origins. After 500 m (1600 ft) of Subway, the splendid tube is lost into the roof of the First Fissure - a deep and narrow vadose stream canyon that was cut into the floor of the older tube after the water table had fallen. First Fissure characterises Castleguard. Too narrow at floor level, too wide at roof level, every caver who ventures in has to traverse on thin, sloping, muddy ledges for 1500 m (5000 ft), frequently

Above: **The ice plug**
The end of Castleguard Cave -
the plug of ice squeezed into the
phreatic tube that lies truncated in
the floor of the basin filled by the
Columbia Icefield.

changing levels to find a passable route. The Grottoes follow, where a level floor provides the only decent camping spot in Castleguard; return trips to the end of the cave take at least three days.

After this comes Second Fissure, another floorless vadose canyon. Worse than First Fissure, its ledges are thinner, steeper and muddier; it rises to include many tricky climbs, some over fearsome gaps, and is 2500 m (8000 ft) long. This awful fissure passage ends at an aven - an upward step that was once a waterfall cascading into the head of its canyon. The first person to climb this was Mike Boon, on an extraordinary solo trip that kept him underground for a week. At the top of the climb, he then walked through a few hundred metres of phreatic tube, before he was stopped by a sparkling wall of ice. He was the first person to see the floor of the Columbia Icefield, about 300 m (1000 ft) beneath its wind-swept surface. That wall of ice was, and still is, extraordinarily beautiful, but reaching it requires nearly nine kilometres of hard caving.

[The story told by Castleguard]

The ice plug at the end of Castleguard Cave completes the picture of a true sub-glacial cave. In uplands around the world, caves were over-run by glaciers during the Pleistocene Ice Ages of the last few million years, and there is debate over just what happened underground during those cold phases. Many cave geologists assume that ice blocked the fissures in the frozen ground, and caves that lay deep enough to escape the big freeze were left totally inactive, dry and sealed off from any erosion or deposition. Others perceive histories of sub-glacial drainage, with seasonal meltwater from ice and snow finding its way down through bedrock fissures to feed streams that continued to flow through caves beneath the permafrost.

Castleguard provides at least some of the answers. It is effectively a living fossil - a modern cave that is a replica of so many caves during the Pleistocene Ice Ages. The ice plug at its end has been squeezed like tooth-paste along perhaps a hundred metres of passage, until it has just stuck. It does not move. It tells us that, when a glacier or icefield sits on top of a cave, everything is frozen up and blocked with ice. The cave survives, but it is static while the ice is there. And that is the story seen in so many of the caves of the Yorkshire Dales and the European Alps - caves that developed and evolved during each Pleistocene interglacial, but just lay dormant during each Ice Age.

The single, long, main passage of Castleguard Cave is probably well over a million years old, as some of its stalagmites have been dated to nearly that age and can only have formed after the tubes were drained to become largely inactive. It pre-dates development of the Columbia Icefield. However, summer meltwater does flow through parts of the cave. Small streams disappear into narrow fissures in the inner part of Castleguard, into another unseen cave that gathers all the drainage at a lower level. In summer, water pours from springs below the entrance, and occasionally backs up to flood the main cave. But erosion by glacial meltwater is very slow, and these passages are still too small to explore. That part of the Castleguard story is destined to remain a mystery.

Below: Second Fissure

Traversing, in the seemingly endless vadose canyon of Castleguard's Second Fissure, is made all the more difficult by mud and silt left on the sloping ledges by floodwaters of times long ago.

Right: **Frozen pools**

Winter-time in the Castleguard entrance passage, and a pool within a rock basin has frozen completely so that its surface has risen into a glistening ice dome.

Mammoth Cave ▮ Kentucky

LOW WOODED PLATEAUS IN THE KENTUCKY heartland just south of the Green River are capped by sandstone, but the intervening streamless valleys and a scatter of large sinkholes indicate the presence of caves. One hillside holds the historical entrance to Mammoth Cave, which leads into a system of caves nearly 600 km (370 miles) long. This enormous network of tunnels now has 15 km (9 miles) of underground trails, established by the Mammoth Cave National Park, with all the rest accessible only to cavers.

For millions of years, rainfall on Kentucky's Sinkhole Plain has all sunk into the limestone and drained down the very gentle dip (less than a degree) northwest to the Green River. These numerous routes have grown into the long sinuous passages that converge and link to form the huge rambling complex of Mammoth Cave. Through this long history, the river has periodically cut down to lower levels; each time, this has allowed the underground drainage to find new lower routes to outlets, abandoning the high-level caves while it formed new low-level caves. So the passage networks are repeated on multiple levels, adding to the huge length of the cave system.

The ancient abandoned trunk routes are magnificent cave passages. Grand vadose canyons (see p.4), 5 m (16 ft) wide and more than twice as high, meander for many kilometres through the limestone. Even longer are the old phreatic tubes (see p.5), many well over 10 m (30 ft) wide. The tributary passages are smaller, but some form complex mazes where they have found different routes along a bedding plane in the limestone, almost like braided channels in a surface river.

There are four main levels of passages within Mammoth Cave, each younger than the one above, and all within a vertical range of little over a hundred metres. Each was formed close to a level of the water table that was dictated by the level of the outlets into the contemporary Green River. Those by-gone levels are found where the trunk caves pass

Opposite: **Collins Avenue**
This spacious vadose canyon was once a major drainage route through the northern part of Mammoth Cave. It was almost filled with sand and clay before being abandoned by its stream; the sand and clay have since been washed down into a younger passage below.

Opposite: **Elliptical tube**
Mammoth Cave's signature passage is the low and wide elliptical tube of Cleaveland Avenue, which winds for kilometres along a single bedding plane. It once carried the main drainage towards the Green River, but is now high and dry with a pathway engineered for visitors to the show cave.

Below: **Roof channels**
Anastomoses are branching channels where the early cave drained along a bedding plane in the limestone, now exposed in the ceiling where the lower bed has fallen away.

downstream from vadose canyons into phreatic tubes - where a stream route became filled to the roof below the resurgence level. The oldest and highest cave levels contain sand sediments that have been dated at 2.3-3.5 million years old, and the actual passages must be even older. This tells a grand story about the evolution of the American landscape. About 1.5 million years ago, the first Ice Age glaciers swept down from Canada and changed the river systems, so that the Ohio River now had a larger catchment and a greater flow. This caused both it and its tributary, the Green River, to cut deeper - and prompted the major new phase of low-level passages in Mammoth Cave.

[The newer streams] The youngest passages in

Mammoth are the many small streamways that drain down to flood-prone tubes at the lowest level. Some of these pass underneath the dry karst valleys, and provide the links between the passages in each plateau. They also include shafts that drain from tiny stream sinks around the edges of the sandstone caprock - vertical features scattered through a cave system dominated by nearly horizontal galleries. Most of the big old passages are kept almost dry beneath the umbrellas of sandstone caprock, so Mammoth Cave has very few calcite stalactites and stalagmites. But tiny amounts of percolation water contain sulphate derived from pyrite within the caprocks, and this is deposited in the cave when the water evaporates, to form gypsum crystals and flowers that richly decorate some passages.

Mammoth Cave is so extraordinarily long because its situation is almost perfect for cave development. Its large catchment supplies ample drainage down the gentle dip to the resurgences, and conspicuous bedding planes provide a host of parallel and continuous routes through the limestone. Successively lowered valley floor resurgences have generated repeated levels of cave passages, of which so many have survived under the protection of the plateaus' strong caprocks. Nearly a thousand kilometres of cave passages are already known in and around Mammoth Cave, and connecting links will eventually be explored to make this incredible underground network even longer.

Lechuguilla Cave ▮ New Mexico

[T H E W O R L D ' S M O S T B E A U T I F U L C A V E]

HIGH ON THE GUADALUPE MOUNTAINS, not far from Carlsbad Caverns (see p.30), an isolated shaft is unrelated to the surrounding topography. It is the top of a small, domed cavern formed from below, and was exposed purely by chance, by erosion lowering the ground surface (see p.5). In 1986, a choke of fallen rocks at the foot of the shaft was cleared away by cavers to reveal passages beyond that now reach a total length of nearly 200 km (125 miles). Named after a locally abundant agave, a plant with fearsome leaf spikes, Lechuguilla Cave is not only very long, but can lay good claim to being the most beautifully decorated cave in the world.

A succession of inclined phreatic tubes (see p.5), deep rifts and dry shafts descends nearly 300 m (1000 ft) to a small junction chamber from where the main galleries extend in all directions, while gaining little more depth. There are sections of tube 5-20 m (15-60 ft) in diameter, which were the main drainage conduits totally flooded beneath water tables of long ago. There are also chambers, each up to 50 m (160 ft) across, scattered throughout the cave. But much of the passage length in Lechuguilla is provided by the incredibly complex and tangled 3-D mazes of passages that are man-sized or not much larger. The most extreme of these are described as boneyards, where the rock has been eaten away so much that its skeletal remains are like a gigantic form of Swiss cheese. The most complex areas of these boneyard labyrinths lace through giant breccia pipes - vertical zones of heavily fractured and mineralised rock.

Unlike most limestone caves, Lechuguilla has no long streamways, no convergence onto a grand trunk passage and almost no roof inlets bringing in extra water from the surface. The overall morphology of the cave makes it clear that the water that formed it arrived from below. Much of it rose up some deep rifts and fissures, which are now choked with debris. The boneyard mazes, the larger passage mazes and the whole network of passages spread out from these source areas - which is the key to understanding how Lechuguilla formed.

The huge abundance of gypsum in the cave then confirms the story that the cave was formed not by rainwater, but by highly corrosive acid.

[Formed by sulphuric acid]

East of Lechuguilla, the deep geological structure of the Delaware Basin contains the source rocks and reservoir rocks of a major oilfield. Perhaps as much as 20 million years ago, the organic debris within these sedimentary rocks was warmed by geothermal heat so that it matured into oil; at the same time, bacterial action on beds of gypsum within the rock sequence produced hydrogen sulphide. Dissolved in water, this sulphide gas migrated through the permeable rocks and rose into the older reef limestones, which already surrounded the basin and now form the Guadalupe Mountains. There it met oxygenated rainwater entering from above, and was oxidised to form sulphuric acid at and below the contemporary water table. This ate away at the limestone to produce the huge networks of cave passages, and its reaction with the bedrock produced the new generation of gypsum that now decorates the caves.

The sulphuric acid reacted with everything. Alteration of the clay beds within the limestone sequence resulted in the formation of new minerals, including alunite. Measurements of the ratios of unstable radioactive isotopes of argon indicate the age of the alunite, which was formed at the same time as the caves. So, it is now known that the main levels of Lechuguilla formed about 5 million years ago, while some of the higher levels in series of passages nearest the entrance had formed about 6 million years ago, before the water table fell in response to surface lowering.

It is difficult to conceive the acid-bath environment in which Lechuguilla was created those millions of years ago. While new passages were being eaten out by acid beneath the water table – perhaps it should be called an acid table – chambers at higher levels were swept by swirling clouds of acidic vapours. Condensation from these created more acid, which was responsible for many of the gypsum decorations within the cave, and also corroded the roof. This upward enlargement of some chambers was important as it was one of them that was later exposed by surface erosion to provide the only known entrance into the cave. Mexico's Cueva de Villa Luz (see p.40) gives just a hint of these conditions, but Lechuguilla lay much deeper and probably in an even more aggressive environment. And its much longer period of acid activity produced the wealth of mineral decorations for which Lechuguilla is rightly famed.

Opposite: **Chandeliers**
Deep within Lechuguilla's maze of passages, the Chandelier Ballroom is a chamber adorned with the largest and most spectacular chandeliers known anywhere - hanging masses of white gypsum crystals that branch like ghostly tree roots and sparkle in the light of a visiting caver.

[A fantasyland of cave decorations]

Calcite and gypsum vie with each other to produce the most beautiful decorations within Lechuguilla. The gypsum came first, mostly created when each cave passage was left just above the falling water table and was still a hot-house of acid vapours. It forms snow-white linings along huge lengths of passages, a relic of the acid reaction with the limestone, and has also accumulated as great thick beds within some chambers. Many gypsum crystals grow from their bases to create either curved flowers or fragile hairs, either as tangled masses or as single strands of gypsum hair that have been found with lengths of up to 5 m (16 ft). Larger crystals form where seepage of sulphate-rich waters feed more mineral to the growth points. The trademark image of Lechuguilla is the sparkling white gypsum chandeliers, curved pendants of branching crystals that reach 6 m (20 ft) long. The finest are in the Chandelier Ballroom where they are fed by dripwater from a thick bed of gypsum within a separate chamber 15 m (50 ft) above the Ballroom ceiling.

After the gypsum, came almost every type of calcite deposit known in caves anywhere. Chambers and galleries contain stalagmites 15 m (50 ft) tall, long stalactites and grand flowstone cascades. But it is the pool deposits that are the most special. Water is scarce in Lechuguilla and many of the rare pools balance their inflow of dripwater with losses entirely by evaporation. This increases the concentration of minerals in solution and causes the deposition of spectacular pool deposits. Grand shelfstone rims (see p.6) are a feature of both active pools and bygone pool sites, and underwater crystals and helictites (see p.6) are both on a magnificent scale. Thin rafts of calcite (known as cave ice) form on the pool surfaces and are held aloft by surface tension, until a drip of water from the cave roof breaks the tension and sinks the raft. Repeated drips from the same points then create cones of rafts on the pool floors. Hoodoo Hall has now lost its cave lake, but slender cones up to 3 m (10 ft) tall are crowded across its floor in a surreal landscape.

Along with pure white aragonite trees (see p.7), dark brown pendants of iron oxide, coatings of bright yellow sulphur, and even bright yellow uranium oxides, the mineral decorations within Lechuguilla are so varied that they have re-written the books on cave minerals, besides creating an incredibly beautiful cave.

Carlsbad Caverns ▌ New Mexico

[W O R L D ' S   L A R G E S T   T O U R I S T   C A V E   ]

Opposite: Lake of the Clouds
At the lowest point known in
Carlsbad Caverns, a small chamber
takes its name from the billowing
clouds of white calcite, whose
rounded shapes are evidence that
they formed when the whole
chamber lay beneath the water table.

OUT IN THE SEMI-DESERT of southern New Mexico, a range of barren highlands is hardly the place where a large cave would be expected. But the Guadalupe Mountains are good strong limestone, and a large hole in one hillside is the single entrance to the very large passages and chambers of Carlsbad Caverns. A well-engineered footpath snakes down a huge bedrock ramp into the open hole, for the cave has been developed for visitor access since the 1920s, and it is still one of the world's truly great underground sights.

The entrance passage descends 60 m (200 ft) into a giant corridor that extends in both directions, about 50 m (160 ft) wide and 30 m (100 ft) tall. The path doubles back under itself and begins a long descent that offers spectacular views down the steeply inclined Main Corridor. Some parts of the roof are massive arches, rounded by dissolution of the limestone long ago when the whole cave was full of gently swirling water. Other parts have angular profiles that trace joints where great chunks of rock have fallen away. The path loops round and partly underneath one huge fallen block that is known as Iceberg Rock; this is over 40 m (130 ft) long and weighs around 100,000 tonnes. It dropped from the roof when it was undermined by water eating away at the bedrock, and it remains a fine example of how a cave has been modified by rock collapse during its long history. Breakdown (p.7) is a part of every cave's evolution, but events are spread so thinly, across the huge span of geological time, that the chances of a caver or visitor being hit by a falling rock are about the same as being hit by a meteorite.

[Big Room] The giant block-strewn passage levels out where it meets the

main level of Carlsbad, just over 200 m (650 ft) below ground surface. A network of spacious tunnels enlarges into a series of grand chambers, including the Big Room. This is really a giant L-shaped passage 50-100 m (160-320 ft) wide for its entire length of nearly 500 m (1600 ft). Most of these tunnels and chambers are less than 30 m (100 ft) high, but huge arches in the roof and wide pits in the floor more than double that height in many places. Perhaps more

Above: Calcite deposits

Stalactites hang down to meet tiered stalagmites that rise from a floor of pool deposits and shelfstone, all formed of golden-yellow calcite in the Chocolate High Passage.

impressive than the sheer size of these caves is their profusion of very large stalactites and stalagmites. The Giant and Twin Domes are three, huge, rounded stalagmites each standing more than 20 m (65 ft) tall within the Big Room. The nearby chamber of the King's Palace is so liberally draped with stalactites that almost no bedrock remains visible. And Lake of the Clouds (in a chamber far beyond the tourist trails) is justly famed for its rounded billows of calcite that were deposited underwater.

For many years Carlsbad's huge chambers were just thought of as an anomaly in the desert environment, perhaps a relic of wetter climates in the distant past. But the soft, porous rock on so many passage walls is unlike the polished limestone typical of stream caves, and the huge thicknesses of bedded gypsum within the big chambers also point to a rather different origin. The evidence came together only in the 1980s, when it was realised that the cave had been largely formed, not by descending rainwater, but by rising sulphuric acid. The theory was confirmed soon after, when Lechuguilla Cave was discovered and studied (see p.27), and has since given rise to much debate over the role of mineral acids in the very early stages of the formation of so many caves.

[Carlsbad's bat colony] It was

the bats that first drew attention to the cave. When seen from afar, the evening bat flight looks like a plume of smoke rising from the cave entrance and trailing off to the east. In fact, it is a spiralling line of countless thousands of Mexican free-tail bats heading out for their nightly orgy of feeding on insects over the nearby Pecos Valley. During the day, the bats hang from the cave roof, forming huge, tightly-packed clusters of furry, twitching bodies along the high-level tunnel opposite the route down to the Big Room. There the bats find the perfect safe shelter for their daytime rest; but there is no food inside the cave. So, like all cave-dwelling bats, they emerge to feed at night, and Carlsbad is one of those sites where they emerge in the huge numbers, making an evening bat flight one of nature's great spectacles.

Local Apaches were probably the first to enter Carlsbad Caverns, but just to take shelter in the daylight zone. It was the 1880s before local cowboys ventured further; they explored along the bat cave, and found huge banks of guano. This was pure bat dung, accumulated over the centuries beneath the roof-top colonies of well-fed bats, where there was no other sediment to be deposited in the cave. Guano is a nitrate-rich fertiliser, much valued in farming. So the cowboys became miners, and dug over 100,000 tonnes of guano from Carlsbad's bat cave, until they ran out of the resource that the bats were replacing at a far lower rate. Many of those cowboy miners were inquisitive enough to explore in the opposite direction, down the giant descending tunnel away from any bats. Among them was Jim White, the leader of the group who turned up a ramp out of the Main Corridor and became the first to set foot in the Big Room. In his spare time he built a rough trail down into Carlsbad's main caverns, so that he could show friends and visitors around. Word spread about the huge discoveries, and it became a National Park in 1930.

Opposite: **The Big Room**
Carlsbad's splendid Big Room houses a host of giant stalagmites; these were formed by dripping rainwater long after the cave itself was eaten out of the limestone by strongly acidic groundwater about three million years ago.

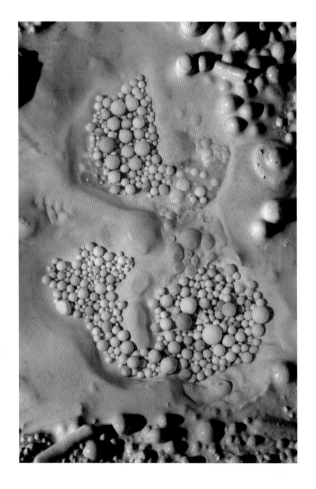

Right: **Cave pearls**
Three small nests of beautiful cave pearls formed in pools whose lime-saturated water was constantly disturbed and replenished by drips from the ceiling of the high-level Rookery Passage; the largest pearls are each over a centimetre across.

For many years, the Big Room was the largest known cave chamber in the world. Larger chambers have been discovered in recent years, but all are accessible only to experienced cavers. The chambers are all younger features with floors that are chaotic heaps of giant limestone blocks fallen from the roof; none of them has a display of giant stalactites and stalagmites to match the splendour of Carlsbad's Big Room. So there is still no show cave experience comparable to the gentle stroll round the perimeter of the Big Room, with its almost level trail more than a kilometre long. Carlsbad Caverns and Skocjanske Jame (see p.104) vie as the world's greatest show cave, offering the choice between a giant river passage and a giant decorated chamber; they are both truly great caves.

Kazumura Cave ▌Hawaii

[THE WORLD'S LONGEST LAVA CAVE]

ABOUT 400 YEARS AGO, a lava flow from Hawaii's Kilauea volcano kept a molten core while its surface cooled down to form a solid crust. When the eruption ceased, the molten lava drained out, and a cave was left in the solid rock. Kazumura is a typical lava cave, except that it is the longest in the world. It reaches from near the crater rim of Kilauea down its eastern flank almost to the coast, with a single main passage 41 km (25 miles) long - creating the opportunity for the world's longest underground walk. This remarkable tunnel descends 1100 m (3600 ft) down the volcano, though nearly all of it lies just a few metres below the surface. It carries no stream, but just a cool breeze - unlike when it was formed in molten lava at a temperature of 1000°C.

Built by rock instead of being dug out by erosion, lava caves are very different from those in limestone. The river of lava that poured from the Kilauea vent lost heat to the atmosphere as it travelled further from its source, so that it developed a cooled, solidified crust. Beneath that rock crust, molten lava still flowed onward, and lost heat only very slowly because it was insulated from the cold world outside. So the lava could flow further before it solidified. Flowing through a tube, lava can extend to great lengths by feeding molten rock out to the front of the flow. This is how volcanoes such as Kilauea are built on such a huge scale.

Much of Kazumura started as a tube full of molten lava, and developed a rounded or elliptical cross-section just like that of a phreatic cave (see p.5). But on the significant slope down the flank of the Kilauea volcano, the hot lava could flow and cascade down the tube like a cave river. It then eroded the tube floor, not by dissolution or abrasion, but by melting the floor rock in a process known as thermal erosion. So some of Kazumura has the shape of a canyon, just like that of a vadose cave (see p.4), especially in the upper part where the lava was hotter because it was nearer to the source vent. Then, as the eruption died down, most of the lava in the cave drained out, to leave a solid floor with lakes, cascades, streams and rapids all frozen into solid basalt rock. Lava falls are up to 12 m (40 ft) high, dropping down into wide

Opposite: **Lava cave**

A splendid section of lava cave over 4 m (13 ft) high and wide, midway along the main passage of Kazumura. When flowing hot lava cut down into the floor, it left rims along the walls, while hot gases re-melted the roof to form a glaze that dripped into a hanging forest of tiny lavatites.

plunge pools, but most of the cave floor is a stream of classic ropey lava - the well-known style of pahoehoe lava, where the smooth skin was wrinkled up by the flow so that it looks like the edge of a coil of coarse rope.

Like a surface river, the flowing Kilauea lava split into channels that later rejoined to create a braided pattern. This survives in the cave with side passages mostly at roof level, abandoned when the main channel was cut down to lower levels as it captured the largest flow. Many side passages end at daylight, and there are now 82 entrances to Kazumura. Most of these were once skylights, where a thin crust had either collapsed or never formed while the stream of molten lava was still flowing beneath. A handful of them are later collapses of the solidified roof, recognisable because the fallen blocks are still there below them, instead of being strewn along the cave by lava that was still flowing when they fell in.

Opposite: **Lava drips**
Knobbly lavamites stand over a metre tall beneath matching lavatites in one of Kilauea's splendid lava caves.

Below: **Perfect tube**
An almost perfect tube in the lower part of Kazumura provides a very long and very dark walk.

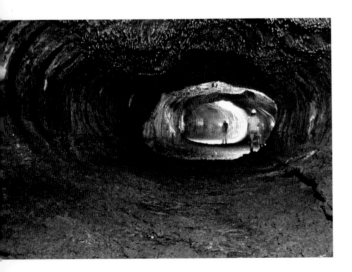

[Decorated by hot gas] When the Kilauea volcano was still erupting, gases that blew through the already well-established Kazumura tube were hot enough to

re-melt the surface of the cave roof. When it cooled, this created a skin of black, glassy lava that now gives parts of the walls a shiny glaze in contrast to the dull, dark grey of much of the solid lava. On the cave ceiling, where the gases were hottest, the glaze was so thick that it dripped off, to form lavatites - pendants rather like stalactites but made of dripping lava that cooled and solidified before it fell to the floor. Temperatures and lava viscosity determined whether these are stumpy cones or delicate tubes. Where molten lava dripped from the ends of the lavatites, it could build up lavamites on the floor - similar to stalagmites but always with knobbly profiles, because they built up as piles of congealed lava blobs. There are few lava stalagmites in Kazumura, but many more stand in nearby caves on Kilauea - neighbours within a suite of truly magnificent lava caves.

39

Cueva de Villa Luz ▮ Mexico

[ACID CAVE 'CRUCIBLE OF LIFE']

ITS COMPLEX LITTLE MAZE OF PASSAGES in limestone reaches only 300 m (1000 ft) end to end, but Villa Luz is a very special cave. The modest cave stream is fed not from the forested karst just 20 m (60 ft) overhead, but from fissures in its floor, from which geothermally warmed, sulphur-rich waters emanate. These waters probably originate from an oilfield about 50 km (30 miles) away in the coastal hills of southern Mexico, and may be typical of the mineralised hydrothermal fluids that have circulated deep within rocks since time began. Cueva de Villa Luz is one of only a handful of caves worldwide that may offer an insight into our planet's very early environments - including where life on Earth began.

Fairly normal within Villa Luz are its cave passages. A warren of half-drained, branching phreatic tubes (see p.5) and fissures has been modified by subsequent breakdown (see p.7) and stream channels. Inlets of highly-acidic water corrode and deeply etch the limestone, until the acidic waters become neutralised. Throughout the cave, the walls and floors have few calcite decorations, but are extensively coated with crystals of gypsum - a reaction product of the sulphuric acid vapours with the bedrock limestone.

Distinctly not normal in Villa Luz is its assemblage of cave life. A small chamber at the back of the cave, with a warm spring in its floor and an atmosphere of too much hydrogen sulphide and carbon dioxide and too little oxygen, contains the best of the snottites. These are the pseudo-stalactites of soft white mucus that hang from the ceiling and drip sulphuric acid from their tips. They are communities of bacteria, comparable to the microbial mats within sulphurous pools in volcanic terrains. They derive their energy, not from photosynthesis in sunlight, but from the heat given out when hydrogen sulphide is oxidised to sulphuric acid. These bacteria are the base of a food chain that extends through mites, midges and tiny insects to numerous spiders and a population of cave-adapted fish. All cave life occupies niches in a hostile environment, but this little chamber, affectionately named Snot Heaven, may show how primitive life could have developed within pools of warm, mineralised water.

Sistema Sac Actun ▌Mexico

[THE WORLD'S LONGEST UNDERWATER CAVES]

THE WORLD'S GREATEST COLLECTION of underwater caves drain out to the east coast of Mexico's Yucatan Peninsula on either side of the ancient Mayan site of Tulum. Drainage from the low karst plateau of the Yucatan feeds out to the sea through a huge network of cave passages that mostly lie just below sea level. Holes in their ceilings appear on the surface as the flooded shafts known as cenotes, and these provide easy access for cave divers into the passages below. The area is a major centre for diving in the warm waters, and explorations of the caves are ongoing. Currently two great systems, Sac Actun and Ox Bel Ha, are vying as the longest explored cave, while their known lengths grow month by month; they each have about 160 km (100 miles) of interconnected passages - all underwater.

Sac Actun was first known as a fine series of large flooded tunnels extending east and west from the Grand Cenote. As the divers have found links to other caves, including the long Nohoch Nah Chich, it has grown into a great network with different parts most easily reached through dozens of cenote lakes. Stalactites and stalagmites cannot be deposited underwater on any large scale, but these caves were left high and dry when world sea levels fell by about 100 m (300 ft) during the Ice Ages of the last few million years, when water was held in the continental ice sheets. Mexico then had a gentle rainfall and no ice, so the caves were invaded by lime-saturated percolation water that deposited forests of stalagmites in the dry tunnels. When the ice sheets melted, sea levels rose, the caves were flooded - and now divers can enjoy floating through these decorated galleries.

Yucatan has the world's most thoroughly explored slice of limestone lying beneath the water table, but its hydrology cannot be regarded as indicative of what lies unseen in every other limestone aquifer. This is because its caves have been formed by a rare combination of major groundwater flow from inland entering the coastal zone, where dissolution of the limestone is enhanced at the interface between fresh and salt waters. Sac Actun and its neighbours stand as remarkable and rather special caves which are both very long and very beautiful.

Opposite: **Underwater** *Cave divers swim through a large tunnel close to the Grand Cenote in the Sistema Sac Actun, gliding over forests of stalagmites formed when sea levels had fallen in response to the growth of Ice Age glaciers on the northern continents, and the caves were not then underwater.*

Quashies River Cave █ Jamaica

A POWERFUL STREAM DROPPING OVER thundering waterfalls, and followed like Ariadne's thread into the depths of a limestone mountain, provides some of the most exciting of cave exploration. Small rivers sinking around the margin of the Cockpit Country in northwest Jamaica have created some magnificent cave passages, and none is finer than that traversed by the Quashies River.

A wide collapse sinkhole offers a scramble down through dense vegetation to a floor 40 m (130 ft) below, where the river pours in through a narrow cleft. On the opposite side, a large

Below: **Low water**
The first waterfall in Quashies River Cave in unusually dry conditions, when its shower drops past clean rock and hardly ripples the surface of the lake below.

cave entrance opens into darkness. Easy walking along a boulder-strewn streamway takes the visiting caver to the first waterfall, with a rope descent of just 5 m (15 ft) from a ledge down to a lake. A short swim, and another ledge, then a rope down past the second waterfall, again only short. Another swim, across a longer and deeper lake, then a careful climb out onto a sloping ledge above a narrow channel where the entire river shoots out into blackness. This shaft is 20 m (65 ft) deep, and a descent in the full force of the waterfall is impossible.

The first cavers to go down Quashies, in 1965, included Pete Livesey, who was also a climber of exceptional talent, and he made a delicate traverse round the wall of the deep shaft. Then, a rope could be tied round a spike of rock for a dry descent; they christened it Little Ape. The rope lands on a ledge beside another lake. It's a short swim across to a rock lip where the river hurtles down the Big Ape waterfall. This shaft is 30 m (100 ft) deep, and requires more delicate climbs out to airy ledges in order to keep the rope away from the roaring cascade. The last section of the descent is magnificent, straight down into the middle of a wide lake. Another swim, and another waterfall, then it's a long swim into darkness and silence, away from the waterfalls and into a tall chamber. The passage beyond is underwater (though it is met again in another adjacent cave), so the caver turns around to repeat the lake swims and waterfall climbs back to daylight.

Above: **High water**
The same first waterfall in the Quashies cave in a state of mild flood, when its roaring cascade churns the lake below and offers the caver a significant chance of being rapidly washed to the bottom of the entire cave system.

Caving is a mixture of sport and adventure, and the succession of swims and rope descents down the aqueous staircase of Quashies River offer the very best of dramatic and exciting underground sport. Few cavers have ever managed to follow in Pete Livesey's acrobatic footsteps, and Quashies remains a rarely visited classic hidden away beneath the lush green forests of the Caribbean island.

Gruta do Janelão ▍Brazil

[WINDOWS INTO A GIANT CAVE]

HIDDEN AWAY IN THE PRISTINE FORESTS of the Peruaçu Nature Reserve, Janelão is one of the world's largest cave passages. It lies in the upper valley of the Sao Francisco River 400 km (250 miles) east of Brasilia. Remote and rarely visited, it is one of the world's great caves. Back in the 1920s, the first explorers walked up the Peruaçu River and found it emerging from a giant tunnel. They walked in, and were astounded to walk through nearly 3 km (2 miles) of giant stream passage; nearly all of it is 60-100 m (200-300 ft) high and wide. They emerged from another huge entrance into the Peruaçu Canyon, which continued for another 10 km (6 miles) with short fragments of giant cave passage along its course. The big cave is distinguished by its four giant skylights, roof collapses that let sunlight pour in, so it was named Gruta do Janelão - Cave of the Windows.

Cave and canyon alike are cut in the Bambui Limestone, a bed of very ancient rock that lies almost horizontal and undisturbed on a geologically very stable block within the South American continent. The Peruaçu River drains off adjacent impermeable rocks and is the main artery of a very fine karst terrain. Few tributary passages join the cave, but inlets of percolation water have created some enormous banks of calcite flowstone within the river passage, while stalactites hanging 30 m (100 ft) from the roof are among the longest in the world.

Nothing is known for certain about the timescale of development for Janelão. Erosion processes are slow in today's dry climate, and the small modern river is no match for the scale of its massive cave passage. It appears that the history of the cave must date back over millions of years. Meanwhile, massive collapse is destroying the cave, albeit almost equally slowly. The cave skylights open into the floors and walls of giant, collapse sinkholes (see p.4), which are large enough to be described as tiankengs (see p.62). Canyons, both upstream and downstream of the cave, also owe at least part of their development to collapse, implying that a much longer cave may once have existed. But no cave is permanent in the world of landscape erosion, and Janelão is perhaps in its declining millennia.

Opposite: Skylights

The magnificent river passage in Gruta do Janelão, lit by natural light between two of the great collapse windows that give the cave its name. A smaller skylight breaks the roof almost directly above the person standing on the sandbank to give scale to the huge cave passage.

Opposite: **Grand river cave**
*Clean walls in the thinly-bedded
limestone sweep, from the darkness
zone, out to the daylight pouring in
through one of the great collapse
windows that lie along Janelão's
splendid river passage.*

[Underground light] Janelão may rank only second to Mulu's Deer Cave

(see p.77) among the world's largest cave passages, but it is probably the more spectacular. This is because its skylights, its giant roof windows, let the daylight pour into its interior. There is enough natural light to let visitors walk through the upstream 1400 m (4500 ft) of passage without a lamp; with both walls and ceiling in plain view, they can really appreciate the full splendour of the gigantic cave. Such a walk is an absolute delight, strolling over gravel banks and wading through shallow lakes where the water is comfortably warm. Even walking is not necessary; a visitor can simply lie back on a gravel bank and watch the daily arc of the sun create endlessly changing plays of light and shadow within the huge cave passage. Perhaps Gruta do Janelão offers a glimpse of what many of the world's great cave chambers have hidden in their barely penetrable darkness.

The input of sunlight also means that trees can grow alongside the cave river beneath each skylight, adding those splashes of green that are rare among the yellow and grey tones of rock architecture in caves. Sunlight provides the energy for plant growth, and plants initiate the food chain for animal life. Animals are fairly minimal in deep and dark caves, but not so in Janelão, particularly where the adjacent dolines are sheltered havens for wildlife. Anything from rats to monkeys to jaguars wander from the forest into the cave, and Janelão's huge twilight zones house countless insects, birds, snakes and smaller mammals. But this multitude of cave life is fortunately lacking any large numbers of bats, and the cave is therefore delightfully free of the awful stench of bat guano that detracts from the ambience of so many large caves in tropical climes.

Janelão's explorers during the 1920s were clearly not the first to visit the cave. Within the smooth limestone of the passage walls in and around the upstream entrance, various alcoves are adorned with huge numbers of wall paintings, depicting animals, men and symbols, all in surprisingly bright colours. Their origins are not known for sure, but dated pigments from comparable artworks in other Brazilian caves suggest an age of about 10,000 years. It is more than likely that these Stone Age visitors to the cave entrance walked on downstream through the long gallery lit by daylight from the great skylights, and they very probably continued through the dark cave beyond with the aid of crude flares. So men may have been walking through Janelão for 10,000 years - and appreciating one of the world's truly beautiful great caves.

Pinega Caves ▍Russia

NOT FAR EAST OF ARKHANGELSK, the port city on Russia's north coast, the endless pine forests just outside the Arctic Circle hide a remarkable group of caves. They lie in the gentle slopes of the Pinega Valley, carved into an extensive bed of gypsum. Though they are notable as a very fine group of gypsum caves, they are now better known for their fabulous ice decorations, which develop each year through the long Russian winter.

Perhaps the most splendid of the Pinega caves are those at Zheleznye Vorota - Iron Gates - a small nature reserve around a dry valley, which cuts through the low gypsum escarpment between white crags known rather fancifully as the Iron Gates. No stream occupies the valley,

Below: **Skating rink**

A shallow pool has become an underground skating rink within Lomonosovskaya, the cave at the heart of Pinega's Iron Gates.

because the drainage lies underground within the adjacent ridge, where Lomonosovskaya is the longest of the caves. The streamways of the main cave are beautiful elliptical tubes, more than walking height and 10 m (30 ft) wide, with gravel floors alternating with long shallow pools. High-level passages mark former routes of the stream and include some complex mazes and a few larger chambers.

Gypsum is not as strong as limestone, and the larger chambers are marked by ongoing roof collapse. Individual beds of gypsum that are cantilevered from the cave walls slowly deform by rock creep, until they curve downwards from chamber ceilings - and eventually break away. Everything happens more quickly in gypsum caves; dissolution of the rock is much faster than in limestone, and collapse is more rapid in the weak rock. There is an air of immediacy in Pinega's gypsum caves, where changes can be seen from one year to the next.

Large stream passages are unusual in gypsum. Most of the world's gypsum caves are drained mazes that formed in slow-moving water beneath the water table. The Pinega caves may have achieved some of their grandeur through erosion by powerful flows of meltwater pouring from ice sheets, during the Ice Ages of the last few million years. Glacial water dissolves gypsum with ease, as it does not depend on the carbon dioxide from plants and soils that is needed to dissolve limestone (where caves are not readily formed by meltwater).

Outside the Iron Gates, Pinega's caves offer more variety in their morphology. Some have shafts dropping into complexes of collapse chambers, and others have single, long stream passages. Golubinskaya is more reminiscent of gypsum caves elsewhere in the world, as it has

Above: **Pillars of ice**

Joints in the gypsum roof have fed fences of icicle pillars that rise above a cascade of ice on one wall of the main passage in Lomonosovskaya; all formed within a few months, they will melt away to reveal only a muddy slope as spring turns to summer.

a series of passages following joints within the bedrock, so that it is distinguished by long straight tunnels with circular or elliptical cross-sections. All the Pinega caves lack stalactites and stalagmites - because once the gypsum is dissolved by groundwater there is virtually nothing to cause its re-precipitation to create the decorations so well known in limestone caves. But Pinega compensates in grand style - with its annual crop of ice decorations.

[Wonderlands of underground ice] Winters at Pinega

Opposite: **Shaft to the ice cave**

A snow-draped shaft drops through beds of less white gypsum into Pekhorovsky Provol, a splendid cave lying just south of the Iron Gates; the cave's chambers are lined with ice crystals, formed in the freezing air that pours down the shaft.

strike a delicate balance of temperatures. There is no permafrost this far south, and the ground remains unfrozen at depths of more than a few metres below the surface. But very cold air pours into the caves through their multiple entrances. So the groundwater that continues to flow creates masses of ice wherever it meets freezing air within the caves. Water dripping from unfrozen rock creates fabulous cascades of giant icicles and glittering ice cascades. Some passages are dry, and others are so far from any open entrance that they remain unfrozen, so not adorned with ice. But any zone of rock fractures, within freezing distance of an entrance, supplies percolation water that creates the most beautiful ice grottoes. The ice builds up though the winter, and then totally melts, to be re-created fresh and clean, and a little bit different, each year. These displays of underground ice may be only ephemeral, but they outshine almost any of their calcite cousins in warmer limestone caves.

Icicle cascades from dripping water are not the only decorations. Water vapour freezes inside some of the caves, and lines the walls and ceilings with a giant variety of hoar frost. With months of uninterrupted growth, sparkling ice crystals reach sizes that are centimetres across. They form interwoven forests of hexagonal plates and spikey blades that frost over the entire rock surfaces. Many crumble and fall under their own weight, so the cave floors gain a carpet of coarsely crystalline snow. It all adds to the fairyland image.

Pond ice is another gem of the Pinega caves, where lakes freeze over to form perfect subterranean skating rinks; except that water remains beneath the surface in what were the deeper pools. Declining flows in the depths of mid-winter cause water levels to fall, so that ice floors lose support; and some collapse to form jumbles of inclined ice slabs that are desperately difficult to cross without crampons and ropes. The Pinega caves have been explored by scientists from Arkhangelsk Geological Institute, and few other people ever get to visit. Access is only by ski in winter; then the caves are at their most beautiful, and always pristine - because each winter brings a brand new collection of their magnificent ice decorations.

Left: **Forest of ice**

The highlight of Lomonosovskaya is the beautiful passage that connects its two entrances; abandoned by the cave's main stream, this gallery is only watered by drips from the roof creating forests of sparkling ice columns in the cold wind that blows between the open entrances.

Krubera ▊ Georgia

[T H E   W O R L D ' S   D E E P E S T   C A V E]

Opposite: **Staircase of shafts**
One more of the countless vertical shafts inside Krubera, where the caver does well to keep away from the cold water.

Below: **Starting down**
An innocuous little shaft in the Ortobalagan Valley on the Arabika Massif gives no hint that it leads to a cave more than two kilometres deep.

DEEPEST OF MANY DEEP POTHOLES within the glaciated Arabika mountains of the western Caucasus, Krubera has been explored to a depth of 2191 m (7190 ft) beneath its single shaft entrance. This is the ultimate vadose drain (see p.4), carrying meltwater from snowfields in its high basin, almost straight down through the limestone. Krubera is not a cave of great chambers or sweeping river passages. Instead it is a twisting canyon of narrow meanders broken by a veritable staircase of shafts. Most of these are short descents, but a handful are larger features, each around 100 m (300 ft) deep. The cave gathers trickles of water from innumerable fissures, so that there is an almost permanent stream cascading down the shafts from the 700 m (2300 ft) level on downwards. Unusually, the passage also splits into various branches; each of these once carried the stream, but each has obstacles in the way of further exploration. Currently, the deepest branch drops into flooded passages where cave divers can only make slow progress.

The exploration of Krubera has been an amazing feat, with expeditions every year since 1999, and still continuing. Teams of cavers, mainly from the Ukraine and Russia, work hard to explore new galleries and survey all that they find. This is not an easy cave, with its tiresome rift traverses, miserable crawls, tight squeezes, endless rope climbs, and even a short, flooded section only passable with mini-scuba gear; there are few places where a caver can walk in Krubera. Throughout all this, the cavers have to haul great loads of equipment - climbing gear, diving gear and camping gear. Nearly three kilometres of rope are needed to equip the shafts. Then there is food and fuel - simply for survival underground, when a trip to prospect new passages more than 2000 m (6500 ft) down normally takes well over a week. The target in Krubera is to reach the main drain that gathers all the Arabika streams and feeds the submarine resurgences (see p.4) that are known to exist on the floor of the Black Sea. But it is quite possible that this grand trunk passage may have lain underwater ever since the Black Sea was filled 7200 years ago. Even without its main drain, Krubera is a remarkable cave explored by remarkable cavers.

Tri Nahacu Cave ▌Iran

SALT IS SO SOLUBLE IN WATER that it can only survive at the surface in a dry desert environment. Even there, thin beds of salt offer little scope for developing karst land-forms. So the salt domes of southern Iran have a notable rarity value in creating mountains of thick salt, some of which contain both spectacular surface karst and also some remarkable caves. The best of the caves so far discovered lie in the Namakdan salt dome on Qeshm Island, off the Iranian coast of the Straits of Hormuz. Tri Nahacu is Czech for Three Naked Men, a name given by the Czech cavers who recently discovered the cave, also known as 3N Cave. It is by far the longest salt cave in the world, with more than 6 km (4 miles) of passages that extend from its upper sinks to a resurgence (see p.4) two kilometres away.

Like any other salt dome, Namakdan has formed where the soft, deformable salt has been squeezed up from its original beds at a depth of some kilometres. It is about 6 km (4 miles) in diameter, and its salt has retained much of its bedded and banded structure, which now stands almost vertically around the edge of the dome. The long cave was formed by the very sporadic rainwater sinking into shafts in the highly-corroded salt mountain. Underground, this drainage headed for a resurgence in a valley cut in from the coast, and was in part guided along the upstanding banding round the edge of the dome.

Processes operate fast in salt. The position of Tri Nahacu, relative to dated marine terraces around the coast of Qeshm, suggests that it is less than 8000 years old, yet it is already a mature cave. A passage that originally had a sawtooth profile, with flooded sections in each of the downward U-bends, has evolved into a graded profile, with a very gentle and almost uniform slope along the entire length of the route of its stream. In the original flooded sections, the salt roof rapidly dissolved and eroded, while silty sediments, that originated as the insoluble part of the impure salt beds, accumulated on the floor so protecting it from further corrosion. The cave is distinguished by magnificent flat ceilings, some of them 30 m (100 ft) wide, formed at maximum flood level. However, sediment has accumulated below,

Opposite: **Banded salt**
Most of the clean ceilings of Tri Nahacu expose dramatic and beautiful sections through the salt, banded with layers of iron-rich red clay, tilted to nearly vertical and locally crumpled into tight folds within the great salt dome.

so that many passages are only about a metre high. Between these low sections, chambers have evolved by progressive collapse of the salt roof. Much of this activity produces typical large blocks of breakdown (see p.7), but some chamber roofs have developed into smoothly arched profiles, where the salt has fallen away in single crystals to create underground mountains of clean, salt gravel beneath. Tri Nahacu now has very long stretches of crawling passage, carrying only a tiny stream in normal conditions, which alternate with grand caverns that are both spacious and welcome.

[Decorations of salt] In Tri Nahacu, everything is made of salt,

including its fabulous arrays of pure white stalactites. These do not have the smooth and vertical shapes of calcite stalactites, but nearly all have bent and twisted shapes that mimic tree roots. This is due to distortion of the salt crystals by rapid outward growth, so that they do not grow solely by deposition from dripping water on their tips. Some have large crystals, approaching the style of the gypsum chandeliers in Lechuguilla Cave (see p.28), but most have more overgrowth to give them less ragged profiles. Forests of these bizarre stalactites, each some metres long, make unforgettable sights in some of the chambers.

Alongside the big white chandeliers and stalactites, Tri Nahacu does have some narrow straw stalactites (see p.6). Many of these are adorned with tiny helictites (see p.6), where outward and upward growth is again controlled by patterns in their salt crystal structures. Stalagmites are a rarity on the floors of Tri Nahacu, but pools along the stream are lined with tiny, sharp crystals of salt. Many of these probably re-form after being destroyed each time the cave floods, but some pools away from the flood routes have had time to grow sparkling walls of the cubic crystals that are typical of salt. Between the crystal pools, the delicate helictites and the clustered stalactites, Tri Nahacu is a beautifully decorated cave that is also a remarkable geological feature. Rapid dissolution and re-deposition of the salt make it one of the world's more dynamic environments - a rarity and a truly great cave.

Opposite: **Salt decorations**
In the great salt cave of Tri Nahacu, a forest of beautiful, pure white, twisted stalactites hangs above sparkling crystal pools. They create a fantasyland of salt that is matched in few limestone caves.

Difeng Dong ▌China

Opposite: **The big tiankeng**
The giant sinkhole of Xiaozhai Tiankeng, which forms the middle entrance to Difeng Dong, seen from the lowest point on its rim. Scale is barely given by the restaurant with a brown roof, beneath the upper cliff and directly above the debris cone in the inner shaft; the footpaths above and below are hidden in the trees.

CHINA HAS MORE LIMESTONE than anywhere else in the world, and is riddled with caves. For centuries the local people have been probing for many kilometres into large walk-in caves, usually in search of water supplies, to extract nitrate deposits (for making gunpowder) or to create tourist sites. But exploration of the deeper and more difficult caves only started in the 1980s, when westerners could join the local people and explore caves for fun, instead of just out of necessity. Already, the karst of southern China has revealed many very large and very spectacular caves. Difeng Dong is one among many in China that belong in a list of the world's great caves, but it does stand out for its splendid river passage and its gigantic entrances.

Difeng lies in the eastern tip of Chongqing province, just south of the famous gorges along the Yangtze River. It is a remote region of forested limestone mountains, which had no road access until quite recently. A village has long stood beside Difeng's middle entrance, but the outside world knew nothing of this, the world's largest sinkhole (see p.4), until the 1990s. It is known as Xiaozhai Tiankeng - which roughly translates as Little Village Sky Hole. The word tiankeng is now used internationally to define the largest of these giant sinkholes, and this is the largest of them all. Xiaozhai Tiankeng is more than 500 m (1500 ft) in diameter, and the highest point on its walls looms an incredible 660 m (2150 ft) above the floor.

This is limestone collapse on a gigantic scale. It has formed by the total roof collapse of a wide chamber, where a very large cave river has been capable of removing the huge volume of fallen limestone. There was probably no single giant roof collapse, but the giant hole evolved though multiple roof and wall failures over perhaps a million years. A footpath now circles round inside the sinkhole to reach the sloping, forested ledge that encircles the inner shaft, and has just a thousand steps to reach that level. From the ledge, a steeper path zig-zags down a forested cone of debris that provides a ramp to the tiankeng floor; this path has another 1800 stone steps.

[The big canyon cave] Xiaozhai Tiankeng drops into the midpoint of

the Difeng river cave, which stretches nearly 10 km (6 miles) between its sink and resurgence (see p.4). The sink lies at the lower end of the amazing Tianjing Gorge (also known as the Great Crack), a narrow canyon, kilometres long and in places only a few metres wide. Access to the cave requires a 180 m (600 ft) rope descent down the canyon wall. Then the cave continues as an underground canyon of comparable scale and style, with vertical walls that rise 100 m (300 ft) sheer above silent black lakes that are interrupted by noisy white cascades. Cavers have to climb along the walls of most of this passage, as the river's current offers only a one-way trip to oblivion. Each year they return to find their ropes and equipment ripped out by wet-season floods, and they have still not traversed the entire passage. Another section of the same tall canyon passage has been explored upstream from the floor of Xiaozhai Tiankeng. But progress has slowed where powerful cascades are flanked by precipitous mud slopes that need time and equipment to traverse in safety.

From the passage just upstream of the tiankeng floor, a tunnel has been driven for 1500 m (5000 ft) to the next valley so that the diverted water can drive a small power station. This was an imaginative project created by local Chinese engineers, who built the path down into the tiankeng but never explored the river cave in either direction. A side benefit is that the cave downstream of the tiankeng now only carries a large flow when everything is in flood. In the dry season, the roaring river is replaced by placid lakes, deep canals and waterfalls that are only showers down clean rock walls. The entire passage is another very tall canyon, and the traverse through to the resurgence is a magnificently aqueous experience. The last cascade drops into the head of a gorge, from where it is a long walk back up to the plateau.

In geological terms, Difeng is just a simple canyon cave following the limestone joints from sink to resurgence as it drops through a limestone plateau, with a shaft entrance midway. But it is on a truly monumental scale, and clearly ranks as one of the world's great caves.

Opposite: **Canyon passage**
The imposing canyon passage of Difeng Dong, looking out of its upstream half into the daylight in Xiaozhai Tiankeng; the water level was very low when this picture was taken, and this section of the river cave is easily traversed because it has a very gentle gradient.

Akiyoshi-do ▌Japan

[B E A U T I F U L   T O U R I S T   C A V E   ]

Opposite: **Stream exit**

Japanese maples provide swathes of red in the autumn foliage around the tall entrance of Akiyoshi-do, which carries the cave's stream out and lets the visitors in to a pathway through a large and beautiful cave.

JAPAN IS NOT RICHLY ENDOWED WITH CAVES, but it does have one little gem tucked away near the western tip of Honshu Island - the very lovely Akiyoshi-do. This splendid cave lies beneath the limestone upland of Akiyoshi-dai. Rolling grassland, pitted by sinkholes and broken by innumerable stone teeth of white limestone, forms a delightful piece of karst terrain among the volcanic landscapes that dominate so much of Japan. Rainfall from a large slice of the Akiyoshi karst drains into the cave, and emerges as a powerful stream from a tall cave entrance in limestone cliffs along the very edge of the karst. Though its entrance is framed by native woodland, Akiyoshi-do is hardly in a wilderness environment, as it has been developed as Japan's premier tourist cave. Winding roads, large car parks, the essential bus parks, well-maintained pathways, a substantial museum and visitor facilities have been created in the immaculate style that characterises Japan's busy tourist sites.

The tall, narrow entrance immediately breaks out into the main feature of the Akiyoshi tourist cave, a grand gallery 30 m (100 ft) wide and varying between 8 and 30 m (25 and 100 ft) tall. The tourist trail follows this large, gently-inclined stream passage for over half a kilometre. The stream winds between banks of sediment and breakdown blocks (see p.7), and the tourist path passes beside sections decorated with very fine calcite deposits. Pride of place among these goes to Hyakumai-zara, a cascade of terraced gour pools (see p.6) fed by a tiny inlet of water, saturated with lime so that it deposits calcite as it flows across the cave. Though each of these gour barriers is less than half a metre high, there are about 500 of them, each holding back a brimful pool that slowly overflows to deposit yet more calcite on its self-built rimstone dam. Each chamber along the stream passage is half filled with a massive slope of breakdown blocks rising towards an arched roof of broken limestone. From the crest of one, a narrower inlet passage takes the tourist trail up to an exit on the plateau, while the main stream emerges from a long gallery with seven flooded sections only accessible to cave divers. With nearly 10 km (6 mile) of known passages, Akiyoshi-do is already a significant cave, and clearly there is more to be discovered in these limestone hills.

Right: **Gour pools**

The terraced gours of Hyakumai-zara are by no means the largest in the world, but they are singularly lovely because every gour is full of water, and the whole scene can be fully appreciated in the cave lighting.

GREAT CAVES OF THE WORLD | ASIA

Tham Hinboun | Laos

[THE GREAT UNDERGROUND BOAT RIDE]

STRADDLING THE INTERNATIONAL BOUNDARY from Khammouan in Laos to Quang Binh in Vietnam, a swathe of limestone has been carved into a magnificent karst terrain distinguished by a number of truly superb river caves. Neither the largest nor the longest of these, Tham Hinboun - the Hinboun River Cave - is perhaps the finest because its entire length of 6 km (4 miles) can be traversed from sink to resurgence (see p.4) by boat. Not just by a small rubber dinghy, but by a comfortable eight-man canoe, driven by a powerful outboard. Small groups of foreign visitors come to enjoy this delightful version of armchair caving, but the Hinboun is also regularly traversed by local villagers who find it an easy route into the cliff-ringed basin that lies beyond the upstream end of the cave.

The rainforests of Southeast Asia provide a perfect cave environment, with substantial rivers draining into the limestone from mountain catchments. In Khammouan, the karst has matured into massifs fringed by tall cliffs that overlook beautiful valleys and closed basins. The massifs are riddled with caves that carry rivers from one basin to the next. Drainage in the upper basin of the Hinboun River found an outlet by following the joints underground through a high ridge of limestone. Water has poured through this for a million years or more, enlarging its route into cave passages 10-20 m (30-60 ft) high and wide. The original cave followed a route up and down the bedrock joints, with the river cascading down ramps between down-loops that were filled with water. Time saw the down-loops evolve upwards, as ceilings were dissolved away and floors were buried by sand and gravel. Meanwhile, up-loops were either eroded away or abandoned as high-level passages. So the river now flows along a gently-graded tunnel right through the mountain.

The result is a delightfully accessible cave river winding between sand banks and through long lakes. Travellers have to get into the water at just four places to drag their boats up short sets of rapids over calcite-cemented cobbles. But the rest of the journey is simple cruising through dark tunnels and looming chambers - caving in a rare style of luxurious comfort.

Opposite: **Underground lakes**
A wide tunnel beckons the traveller underground at the resurgence end of the Hinboun River Cave. The villagers' fishtail-powered canoe proves invaluable, as deep lakes floor much of the passage through to the far end of this splendid cave.

Perak Tong ▮ Malaysia

[BUDDHIST HOLY CAVE]

Opposite: **Painted cave**
The main temple cave in Perak Tong, dominated by its golden Buddha that rises over 12 m (40 ft). Its limestone roof is eroded into rounded scoops and hollows now adorned with paintings, each of which is its own splendid work of art.

IN LIMESTONE HILLS down the west coast of the Malayan peninsula, a handful of caves are more important for their human role than they are as natural features. Rising above alluvial plains, isolated hills are remnants of an ancient karst terrain, where natural drainage formed long cave systems of large phreatic passages (see p.5). Today's main caves are relics of very long cycles of erosion, mere segments of those ancient caves, now truncated and drained above the modern water table, so that they are dry except for a little dripwater. Some of the larger caves have become major sites of religious pilgrimage. Well known are the Hindu cave temples at Batu, near Kuala Lumpur, but the Buddhist cave temple at Perak Tong, just north of Ipoh, is even more spectacular.

Within the limestone hill of Gunung Tasek, Perak Tong Temple is built in and around the cave now known as Perak Tong. The cave's passages have walls and ceilings, rounded into alcoves and domes, that are classic features of caverns carved out by slow-moving water at a time when the rock was far below the water table. They honeycomb the hill, with a main entrance just above the plain and behind a traditional temple, and exits right on top of the hill.

Perak Tong's entrance chamber is a classic cave shrine, developed in 1926 by a Buddhist priest from China, and dominated by a splendid golden statue of the sitting Buddha. But what really distinguishes Perak Tong is the huge collection of religious paintings all over the cave walls and ceilings. Carved by swirling water, the smooth limestone has almost no stalactites to spoil rock surfaces that are ideal for mural paintings. The cave temple now doubles as a major centre of Buddhist art, and the finest artists from the Buddhist world have been invited to add their own murals. Each is in its artist's own style, and the variety and sheer number of paintings make the cave an extraordinary place to visit. Perak Tong may lack the antiquity of the cave paintings at Chauvet (see p.92), but the whole cave is beautiful. A painting of the smiling and very rounded Chinese monk popularly known as Fat Buddha, finds a perfect setting on an equally rounded wall pocket carved into the wall by ancient groundwater.

Caves of Mulu ▮ Sarawak

[THE WORLD'S LARGEST CAVES]

Opposite: **Sarawak Chamber**
Only with giant flash-guns held by cavers can the awesome rock architecture of Sarawak Chamber be seen through the blackness.

Below: **The Pinnacles**
Mulu's mountains are etched into an impossible terrain of razor-sharp limestone, none more spectacular than The Pinnacles that rise 30 m (100 ft) through the forest canopy on Gunung Api.

IN THE 1970s, the rainforests that clad the sandstone mountain of Gunung Mulu, in the northern district of Sarawak, were selected for conservation and protection from Malaysia's destructive logging industry. On the western flank of Mulu and included within the conservation zone, a line of jagged mountains of limestone were known to contain some caves. So a project to document the forest and plan its future included a few cavers to see what lay underground. They were more than delighted when they found a series of enormous and fabulous caves. Explorations continue to this day, and more than 350 km (220 miles) of cave passages have already been mapped. Mulu is now a National Park, and a few of the accessible caves are major visitor attractions.

The caves of Mulu are so large because they lie in the perfect cave-forming environment. It rains every day, and the forest creates run-off with plenty of carbon dioxide to maximise the rate of dissolution of the limestone, while large rivers and streams drain from Gunung Mulu onto the limestone outcrops. On top of this, Mulu's limestone is a strong rock, with fractures widely spaced across beds tens of metres thick. This is critical to the existence of the world's largest cave chamber - 700 m (2300 ft) long and 300-400 m (1000-1300 ft) wide - in the cave of Lubang Nasib Bagus. Sarawak Chamber was formed by a cave river that initially followed a single, dipping, bedding plane and slipped sideways down the dip to create the huge area of the chamber. Blocks then fell away from the roof, until a stable arched profile was achieved. There is ongoing debate over the maximum size a cave chamber can reach before collapsing, and this enormous black space inside Mulu's mountain almost defies the imagination.

GREAT CAVES OF THE WORLD | ASIA

Nasib Bagus, with its giant chamber, lies under the eastern side of Gunung Api, the largest of the high limestone mountains at Mulu. Most of the rest of this mountain is underlain by the network of passages that constitute Clearwater Cave (Gua Air Jernih in the local language), named after its river of clear water that pours into the perpetually muddy river system of the rainforest. Its known passages total more than 150 km (95 miles) in length, and more are discovered and mapped each year. Though this only ranks as the world's number eleven, Clearwater has a greater importance because such a large proportion of its passages are very large; it may well be the world's largest cave by total volume. The main river passage is 30 m (100 ft) high and wide for many kilometres of its length, and the main high-levels are of equal size. Each major passage in Clearwater began life as a partially or totally flooded tunnel, guided through the limestone along a steeply-inclined bedding plane at a level close to that of the adjacent Melinau Plain. While the mountains of strong limestone have remained barely eroded, steady lowering of the plain by river erosion has, over a total period of millions of years, allowed a sequence of parallel passages to develop at ever lower levels.

[**Deer cave**] One of the first caves to be found in Mulu was Deer Cave, because its entrance was unmissable - it is well over 100 m (300 ft) high and wide. This massive size continues right through the mountain to the sink (see p.4) over a kilometre away - the largest cave passage in the world. It was formed by a major river draining along the limestone, parallel to the Clearwater passages, for an unknown but very long period of time.

Though its upstream end is in a valley that was formed by major collapse, there are relatively few breakdown blocks (see p.7) within its one giant passage. Instead, Deer Cave is known for another type of debris on its floor - the disgustingly aromatic bat guano. A colony of some few million bats lives in the cave. Every evening there is a mass exodus in a spectacular bat flight, as they leave to feed on the insects of the rainforest. But by day they hang from the cave roof 120 m (400 ft) above the floor in a tightly packed and heaving mass, and their droppings form a veritable rain of organic debris. Huge quantities of this stinking bat guano have accumulated, with one pile that is over 30 m (100 ft) high. And this is the base of the food chain for a whole host of creatures - the cave is alive with millions of earwigs, beetles, cockroaches and crickets together with the giant spiders and centipedes that are their predators. This seething mass of cave life is just one more spectacular feature of Mulu - a special place that is perhaps the ultimate in cave development.

Opposite: **Deer Cave**
The enormous tunnel has daylight reaching in for half a kilometre. Only a small stream now meanders across the floor, where a powerful river once flowed, while a few million bats hanging from the roof contribute their own variety of malodorous sediment to the cave environment.

Left: **Api Chamber**

In the largest chamber within Clearwater Cave, the limestone bedding is just visible dipping steeply to the right. Breakdown blocks are piled on the floor because the river that formed this ancient cavern now pours through a cave at a lower level.

Naré ▌ New Guinea

OFF THE NEW GUINEA MAINLAND, the smaller island of New Britain is remarkable for its great karst plateau of Nakanai. Clad in thick forest and doused by a huge annual rainfall, Nakanai is drained by huge underground rivers, some of which can be entered by enormous shafts. It is one of the most remarkable and spectacular karst terrains in the world. Naré is not the longest or deepest of the Nakanai caves, but it is the finest. Its huge shaft into a massive river passage is perhaps the archetypal cave, whose exploration is a monumental challenge, most of which was first achieved by teams of French cavers. Though its passages are so large, Naré is a youthful cave formed by a river draining directly along a dipping limestone bed; its size is simply due to powerful erosion fed by the huge rainfall. The entrance shaft then formed where progressive collapse of the cave roof met the eroding floor of a deep sinkhole. That too was driven by rainfall eating into the rock.

Opposite: **Ultimate entrance**
A single rope takes the caver down more than 200 m (700 ft) into the booming river passage of Naré.

Below: **River crossing**
Nare's underground river can only be crossed on a tyrolean - a tensioned rope that has to be first established with a grappling iron to gain a hold on the far side.

Naré's shaft is almost circular at about 100 m (300 ft) in diameter. It is the classic hole in the ground, and it is 230 m (750 ft) deep, descended by cavers on a single, unbroken, free-hanging rope. The abseil descent gives wonderful views of the huge river tunnel entering on one side and leaving on the other. From the bottom of the rope, the river passage descends only gently, but this cave is no easy walk. The river is large enough and fast enough to sweep anyone to an aqueous demise, so the cave can only be traversed along ledges and boulder banks along its sides. Easy for short stretches, except for hazardous tyrolean crossings necessary where the river swings from one vertical wall to another. Little more than two kilometres from the shaft, progress ends at an impassable tunnel nearly full of roaring water, but Naré is already a truly great cave.

Nullarbor Caves | Australia

[TUNNELS UNDER THE DESERT]

Opposite: **Grand tunnel**

The grand tunnel in Abrakurrie Cave has a beautifully smoothed roof arching through the almost pure white limestone. It sits silent and dry, deep beneath the Nullarbor Desert, except when a rare rainstorm floods it with muddy water to leave the head-high tidemark.

AGAINST THE SOUTHERN OCEAN COASTLINE, the vast expanse of the Nullarbor Desert is a flat plain formed on soft, flat-lying limestones. It has few surface karst features, but it hides a dozen or so rather grand caves that are distinguished by their spacious dimensions. The single passage in Abrakurrie Cave is more than 20 m (60 ft) high and wide, a short fragment of an ancient tunnel that is now choked with sediment at both ends. Time-scales in Australia are huge, and these caves largely formed as flooded tunnels in wetter climates millions of years ago, when drainage from inland sandstone outcrops found underground routes to the ocean. Now, there is little flow through the desert limestone, and the groundwater stands at lower levels. Abrakurrie is dry (except after rare storms), Weebubbie has a long lake at the water table, and Cocklebiddy has a long tunnel still totally flooded.

Today's erosion processes in the Nullarbor caves are unlike those anywhere else. The minimal desert rainfall is rich in salts because it is derived from sea spray. As it percolates down through the porous limestone, it picks up calcite by dissolution, but retains its salts - until it reaches the ceiling of a large, dry cave. There, evaporation causes precipitation of the salts within the limestone's pore spaces, and crystal growth just breaks the rock apart. This granular breakdown is known as crystal weathering, and is on such a fine scale that it creates smoothly rounded wall profiles that mimic the shapes produced by underwater dissolution. Salt, sodium chloride, is the main mineral behind this process, and some of it is also deposited within the open caves to form delicate stalactites and crystal flowers.

Crystal weathering has created the smooth roof outlines of the Nullarbor caves, and also their floor sediments, but this is locally interrupted by rock collapse on a larger scale. Koonalda is the cave best known for its splendid roof domes that soar above matching piles of breakdown (see p.7), but many of the caves only have accessible entrances where roof failure broke through to the surface. Though few in number, large collapse sinkholes

characterise the Nullarbor karst. The huge sinkhole at Abrakurrie's entrance is just one in a line dotted across 10 km (6 miles) of the desert, suggesting that its short passage is only a tiny segment of a long and large tunnel, unknown until a way can be found through the fallen breakdown debris in the other sinkholes.

Unseen caves far outnumber the known caves under the Nullarbor. Beside the collapse sink-holes, blowholes are the only other evidence of their existence. Caves breathe, because the amount of air inside them varies in response to pressure changes induced by passing weather systems. When pressure declines in the face of a storm, air flows out of the caves, only to flow back in as pressure rises again. A small entrance to a large cave creates a powerful blowhole with reversing winds that can blow a visitor's hat off. The Nullarbor has hundreds of blowholes, nearly all too small to enter. Some link to known caves, but others merely indicate that there is a lot more cave passage still to be discovered beneath the Australian desert.

Left: **Clear lake**

Most of the known length of Weebubbie Cave is occupied by a beautiful lake where the ancient passage meets the water table almost at sea level. Beneath the crystal-clear water, large blocks of limestone lie where they fell, as successive beds broke away from the roof arch.

Waitomo Caves ∎ New Zealand

Opposite: **Silken traps**
Fishing lines from the glow-worm larvae hang from the roof of the Mangawhitikau Cave in the Waitomo limestone karst; their beads of sticky mucus catch the light, but the glow from the larvae themselves only shows up in total darkness.

FOR ANY ANIMAL LIFE IS NOT EASY inside a cave, as there is a shortage of food - because there is no sunlight to encourage growth of plants as the first link in the food chain. Each permanent cave dweller therefore has to adapt to the underground. Few are more cleverly adapted than the larvae of an inconspicuous little fungus gnat that thrives in the caves of the Waitomo area in the North Island of New Zealand. They are renowned because they use light to lure their prey, and are popularly known as glow-worms.

This particular gnat, *Arachnocampa luminosa*, glues its eggs onto the cave roof. When these hatch into larval grubs, they weave little nests of silk also attached to the roof. From these, they hang multiple fishing lines of silk armed with tiny beads of sticky mucus. Clouds of beaded lines, perhaps 40 cm (15 in) long, hang from thousands of cave nests within a single cave. In each nest, the larva, up to a few centimetres long, waits with its tail segment glowing in the dark. Midges, gnats and flies are the prey; they follow the stream into the cave, head up towards the forest of lights, and are caught in the sticky lines. Hauled up on the lines, they are soon consumed, but each larva keeps its tail-light going to attract the next meal.

The glow-worms live in cave chambers that are traversed by streams not far in from the open air. The Glow-worm Grotto, just 30 m (100 ft) in from daylight in Waitomo Cave, is the best-known; since 1889, it has been open to tourists who take a boat ride across the cave lake to gaze up at the myriad specks of glow-worm light across its roof. Truly memorable, but the chamber is not unique, as the glow-worms will happily hang from almost any cave roof that has a stream below and an entrance nearby. The Mangawhitikau Cave is another glow-worm site accessible to tourist visitors, and cavers can reach many other caverns inhabited by glow-worms throughout the limestone hills of the Waitomo karst. Where an animal finds the right niche in the cave environment, it takes up in huge numbers, and the Waitomo caves have proved just right for the glowing gnat larvae that interrupt the total blackness of the underground with a magical starlight.

Gaping Gill ▌England

Opposite: **Waterfall**

Flood conditions turn the main chamber of Gaping Gill into a maelstrom of crashing water. All the water just disappears into the deep floor sediments.

THERE IS SOMETHING SPLENDIDLY MYSTERIOUS about a large mountain stream that suddenly drops into oblivion down an unfathomable shaft. High on the slopes of Ingleborough, in the glaciated limestone country of the Yorkshire Dales, Fell Beck does just that, and Gaping Gill is a really good name for the big, black hole that swallows the entire stream. Only when cavers went down the plummeting pothole of Gaping Gill did they find even more drama below. The great vertical shaft opens out into the roof of a grand chamber, with an almost flat floor of stream cobbles 100 m (320 ft) below moor level. Fell Beck drops nearly into the middle of the chamber as an unbroken waterfall, with the whole scene dimly lit by daylight from above.

Piercing the walls of the chamber are three dry tunnels, each of which offers entry to a rambling cave system with more than 16 km (10 miles) of mapped passages. Most of these are very old tunnels, now dry, though their rounded profiles show that they were totally flooded when formed by the water draining away from the main chamber. This network of tunnels follows the bedding planes in the limestone, so it is almost level except where it steps down to lower beds. Infamous among these is Hensler's Passage, which is 400 m (1300 ft) long and nowhere more than a metre high; it is the ultimate bedding plane cave that lost its stream long before it was enlarged to a more spacious tube or a deep canyon.

Small streams have invaded this network of ancient tunnels. They pour in from more than a dozen entrances on the rain-washed moors, through splendid little cave systems of steep canyons with staircases of waterfall shafts that eventually drop into the older tunnels. There is also a route out to Clapham Beck Head, where Fell Beck returns to daylight from a very modest cave mouth. The lower end of this route has been a show cave since 1838, soon after a calcite barrier was broken down to drain a lake and reveal the stalagmite-decorated passage now known as Ingleborough Cave. But the rest of the route is inordinately complex as it follows through different sections of old passages that are partly choked by sediment.

GREAT CAVES OF THE WORLD **|** EUROPE

[Still the unknown]

Gaping Gill guards its secrets closely. Fell Beck sinks into the cobble floor of the main chamber, and almost none of its route is known until it re-appears in the back end of Ingleborough Cave. Equally mysterious are the large, old, dry passages that radiate from the chamber and end abruptly in very solid chokes of sediment and rock debris. It is clear that these old passages lay for millennia far beneath the glaciers that repeatedly covered the Dales during the Ice Ages of the last few million years, when meltwater washed debris into their many entrances.

Even more mysterious is the main chamber. Recent investigations with ground-probing radar appear to indicate that the cavern's floor sediments are at least 30 m (100 ft) deep. Most of this is stream debris filling a chamber that was once twice the size of what can be seen today. And deep down beneath that pile of sediment lies a flooded tunnel that still carries Fell Beck, far out of reach of cavers and divers alike. Gaping Gill is no longer the deepest shaft or longest cave known in Britain, but it may one day prove to be the oldest. Meanwhile, it is very appropriate that a great cave should hide so many secrets.

Opposite: **Pavements**

The great limestone pavements that characterise Ingleborough are on single bedding planes that were scoured clean by Ice Age glaciers. Rainwater has etched out the joints to form networks of deep fissures that drain down to the caves below.

Below: **Hensler's Passage**

Far below the limestone pavements, long-gone flows of groundwater carved out a parallel bedding plane to form a long and very low cave.

Grotte Chauvet █ France

[S T O N E   A G E   A R T   G A L L E R Y   ]

ALONG ITS COURSE from the Central Massif down to the Rhone Valley, the River Ardèche cuts through a range of limestone hills in picturesque gorges whose rocky walls expose a number of caves. Best known is the Pont d'Arc, where the river glides beneath a high natural arch right through a rock spur. But close by there is another cave of far greater significance. The Grotte Chauvet is only a few hundred metres long, a mere fragment of an ancient cave system, blocked by stalagmites at one end and by cliff breakdown (see p.7) at the other. Its entrance lies on a ledge high in the cliff, bounding the dry valley that loops round the Pont d'Arc. It was only found in 1994, when Jean-Marie Chauvet and two friends dug out some rubble, and squeezed between massive blocks to penetrate the cliff's apron of rockfall that had blocked the cave for thousands of years. Then they found they were not the first to venture inside the hill.

The three friends had reached a cave chamber 40 m (130 ft) wide and over 60 m (200 ft) long. Its clay floor is littered with bones, including dozens of skulls of cave bears, and is pitted with hollows that the bears had dug before settling in for their winter hibernations. At the far end, the roof lowers to little over walking height, and then rises again into a second chamber, rather smaller than the first. Both have walls and ceilings decorated with sparkling calcite stalactites. The three explorers had only small helmet lamps, and it was only on their way out that they caught sight of a pendant of rock with the figure of a mammoth drawn on it in red ochre. They looked around more carefully and found more and more wall paintings. On further visits to the cave, they were joined by teams of archaeologists, and yet more paintings were identified; a total of 420 animal figures have now been recorded in the fabulous underground art gallery that is the Grotte Chauvet.

Paintings in the large main chamber are distinguished by their simple artwork in red ochre - a soft iron oxide mineral that can be found in the limestone hills. There are some beautifully drawn outlines of bears, and also a few hand stencils; these were made by the artist blowing

Opposite: **Horses in charcoal**
The beautiful charcoal drawings of horses heads in Chauvet's Hillaire Chamber are almost brought into relief by the clever shading, which is a feature of this sophisticated artistry from many thousands of years ago.

ochre over a hand pressed to the wall, and are common in painted caves. More unusual are the large ochre spots, each made with the palm of a hand; some are grouped in the shape of animal silhouettes, but others lack any structure.

The finest of the art is further into the Chauvet cave, in the Hillaire Chamber and in the small End Chamber leading off it. Ancient floods had left a thin coating of clay on the walls, so that the cave artists could simply use their fingers to trace the outlines of animals; very simple, but there is superb artistry in the many profiles of horses. Superimposed on these are parallel scratches left by bears standing against the walls, probably to stretch on awakening from their hibernation. On other parts of the walls, the clay layer was scraped away before the artists drew their figures in charcoal straight onto the limestone. Horses, rhinos, deer and bison dominate, and clever use of shading makes many of the horse figures especially magnificent. The artwork in the End Chamber is distinguished by the animal profiles crowded over each other, as if to represent herds of the animals; they include horses, rhinos, lions, ibex and mammoths, and the total effect is spectacular.

Below: **Pont d'Arc**

A highlight of the Ardèche Gorges is the Pont d'Arc, the towering rock arch that spans a very short natural cave carrying the river through a limestone ridge.

The big question is just how old is the cave art. Charcoal can be dated by analyses of its carbon isotopes, whose ratios change over time due to radioactive decay when hidden away from the restorative powers of solar rays. The Chauvet material has been dated to about 35,000 years old. This was a surprising result, because it placed the Chauvet art at nearly twice as old as the well-known paintings in Lascaux and other caves in western France, and Spain. Chauvet achieved instant fame, but serious doubts have since been raised over the validity of these dates; carbon dating can produce significant errors and also requires rigorous interpretation, so there is ample scope for debate. The current view is that

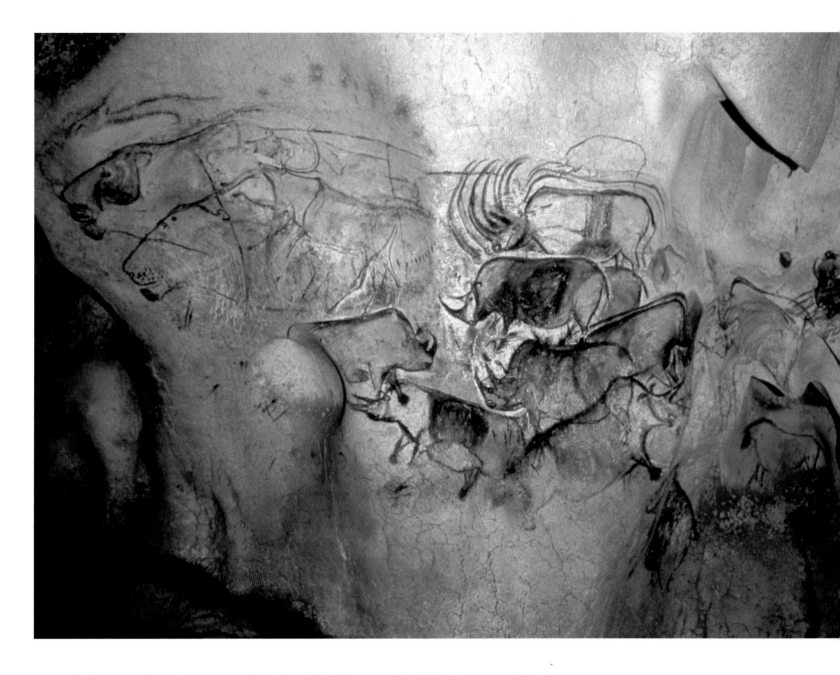

most of the red ochre drawings are less than 30,000 years old, while the more refined artwork in the inner chambers probably dates to 15-20,000 years old, where it broadly equates with the ages and the styles of the Lascaux paintings. The age debate continues unresolved, but the paintings in the Grotte Chauvet are nevertheless of truly fabulous quality, and they remain a glorious testament to the skills of our artistic ancestors who found a special use for caves.

Above: **Lions and rhinos**
Lions and rhinos adorn one wall of End Chamber, both of which are relatively unusual in European cave art, and here overlie the remains of older paintings in red ochre.

Gouffre Berger ▮ France

Opposite: **Hall of the Thirteen**
The splendid cave chamber, seen from the edge of the boulder pile of the Great Rubble Heap. This is the archetypal scene within the Berger, lying nearly midway along the most delightful of journeys of underground sport.

BACK IN 1956, the Berger was the first cave in the world to be explored to a depth of more than 1000 m (3280 ft). Today, its depth is surpassed by 27 other caves, but the Berger is still regarded as one of the world's finest. It is the perfect cave, with a sequence of entrance shafts that drop into a huge river passage, which can be followed into the depths without any horrible narrow bits. No cave this deep is easy, but the Berger is not too arduous; and it is supremely enjoyable.

On a northern corner of the Vercors Mountains, the limestone pavements (see p.90) of the Sornin Plateau overlook the city of Grenoble. Local cavers were looking for the source of the water pouring from the Sassenage resurgence on the edge of their city, and they felt sure the cave had to originate on the Sornin. They searched dozens of fissures and shafts that led nowhere, until Jo Berger found the entrance that gave them the way in. Five shafts, each about 30 m (100 ft) deep, are separated by tall, meandering canyons, which they had to traverse on narrow ledges - a sporting challenge that still demands some effort and makes a visit to the cave beyond even more sweet. Round a corner from the last shaft, just 250 m (800 ft) beneath the Sornin, the narrow canyon breaks into a huge gallery. Jo Berger and his friends had hit the jackpot; they called it La Rivière sans Ètoiles - the Starless River.

The relatively small canyons and shafts from the entrance drop through a sequence of strong white limestones, known as the Urgonian. The much larger main gallery is formed almost along the boundary between the Urgonian limestones and an underlying limestone that is a little more impure and rather more fractured. More rapid erosion of these underlying limestones has cut the large main passage by undermining the strong upper beds, which now form a stable roof across its wide chambers. Just beyond the inlet from the entrance, the geological structure turns down into a syncline, so that the Berger cave plunges to greater depths by following the dip of the limestone; it is a perfect example of geological control over cave morphology.

A short way down the Starless River, Lake Cadoux lies across the whole passage width, and cavers normally paddle across in rubber dinghies. After the boat trip, there is a stroll between the giant stalagmites of Bourgin Hall, before the view ahead opens into a giant blackness, where the passage widens and tips downwards into the Great Rubble Heap. This is a huge chaos of house-sized boulders, where the cavers' expression 'ants in a sugar-bowl' was first coined. It ends in the splendour of the Hall of the Thirteen, where giant stalagmites stand sentinel over a cascade of white gour pools (see p.6). A rocky terrace is a popular site for the first underground camp, and there is something magical about waking in the morning to total blackness, reaching out for a headlight and the first thing in sight is this gallery of stalagmites. The journey continues down a floor of calcite in the giant passage to a balcony, which drops into the canals. There, a narrower slot in the floor provides grand sport for the caver who has to traverse along ledges above deep water and climb down waterfalls. Then the slot opens up into another huge boulder-strewn chamber, but the way beyond involves even more aqueous sport.

Opposite: **Grand Cascade**
A caver abseils down a rope beside the Grand Cascade, one of the waterfalls in the lower section of the Berger.

Below: **Starless River**
Only the photographer's powerful flashlights bring into view the fissured roof of the Starless River - the Berger's main gallery that leads to the depths.

[Down the waterfalls] Each of a

series of splendid cascades drops into wide plunge pools, where airy traverses and delicate climbs are necessary to keep the caver out of the torrential cold water. The last of these is the Hurricane Shaft, named after the turmoil of spray whipped up by the waterfall that crashes 50 m (160 ft) into blackness. This drops into a major joint zone that turns the cave eastwards towards the Sassenage resurgence (see p.4). But the cave also levels out here, so that it runs back into the Urgonian limestone, where its passages are much smaller. The water cascades through narrow canyons and dark lakes that require swimming, until the roof dips down into the water in a sump pool (see p.4). Only the cave divers can go further, but they have reached only a short complex of small passages where the water is lost, and the way on to Sassenage remains undiscovered. It is appropriate for a great cave to end with a mystery. The Berger has all that is the best in a great cave, and the journey to its final pool remains one of the world's great underground adventures.

Alpine Ice Caves ▌Austria

[UNDERGROUND GLACIERS]

STRADDLING THE HEART OF AUSTRIA, the Northern Calcareous Alps are a chain of glaciated limestone mountains 500 km (300 miles), which are famed both for their expanses of corroded limestone pavement and also for their collection of very long and very deep caves. These caves are classic alpine systems with arteries that are narrow, twisting canyons with deep shafts sprayed by very cold water. Many of the shaft systems drop into ancient tunnels that have been left high and dry; this is since the water drained out into adjacent valleys scoured deeper by glaciers of the Ice Ages in the last few million years. Some of these abandoned caves even have their own glaciers inside them. Two of them, the Eisriesenwelt and the Dachstein Eishöhle, have been developed as grand show caves that can give visitors the opportunity to marvel at these fantasy worlds of underground ice.

[Eisriesenwelt]

The Eisriesenwelt lies within the plateau of the Tennengebirge, more than a kilometre above the adjacent floor of the Salzburg valley. Ice Age glaciers cut into the limestone and exposed the older cave passage in a cliff face, leaving the dramatic circular entrance that now leads into more than 40 km (25 miles) of ancient passages. These were once the arteries of a huge, flooded cave system that drained the whole plateau, but rainfall and snowmelt now find their way down new shafts to lower levels, leaving these tunnels high and dry. The main passage is all 10-20 m (30-60 ft) high and wide, and rises only gently along its route into the mountain.

Very special conditions are required for ice to accumulate inside caves. Winter temperatures are low enough in the mountains to freeze the caves, but unfrozen water has to reach the passages during some of the year to feed renewed growth of the ice. When the ground thaws out in summer, water can reach the cave, which has to stay cold enough to freeze the incoming dripwater and then retain its ice. The Eisriesenwelt is linked to the plateau above by fissures, too narrow for cavers, but effective at letting air through. In summer,

the high plateau keeps many of its snowfields, so freezing air sinks through the cave to keep it cold. In winter the cave is warmer than the outside, so the flow of air is reversed, and a cold wind blows in from the lower entrance, but even that is freezing in the alpine winter. What most distinguishes the Eisriesenwelt is the massive scale of its ice. There is very little summer melting, so the ice has accumulated into hugely thick sheets - veritable underground glaciers that creep very slowly within the cave. The first kilometre of the cave is richly adorned with ice, and the first half of that is floored entirely by ice that is up to 10 m (30 ft) thick. Dripwater entering through the cave roof freezes into massive ice stalagmites and columns. These rise high over the main mass of floor ice, some of which is banded just like any glacier in the great outdoors. Walls of ice that are left beside sections of open passage are beautifully carved into large, wavy hollows - larger versions of the scallops so common on limestone cave walls. Instead of forming by rock dissolution in eddies in the cave streams, these have developed by melting and direct evaporation of the ice in air eddies within the gentle cave winds.

[Dachstein Eishöhle]

Neighbours to the east of the Tennengebirge, the Dachstein mountains are also riddled with huge cave systems. High in the limestone ramparts above the Hallstatt Lake, one segment of abandoned ancient passage is known as the Ice Hole because it is half-filled, and richly decorated with grand and beautiful ice. Parts of the cave have walls of clean, dry rock; the main passages follow the gentle dip of the bedded limestone, and their rounded profiles show they were formed long ago when completely full of water. The cave was abandoned when its water drained out into valleys newly deepened by Ice Age glaciers. Water percolated down through the limestone and dripped into the cave, but dissolution rates were very low in the cold mountain environment, and no calcite deposits formed. Instead, the dripping water froze in the cold air that blew gently through the cave. Over about the last 500 years, during and since the medieval Little Ice Age, ice deposits have grown on a magnificent scale.

The Dachstein's underground ice is truly spectacular. Columns, curtains and frozen cascades are among a huge variety of beautiful ice decorations. Just as in the

Below: **Frozen cascade**
In the Parzival Chamber within the Dachstein Eishöhle seeping water, from a bedding plane in the limestone, feeds a frozen cascade of icicles and curtains that fall to a massive sheet of underground ice.

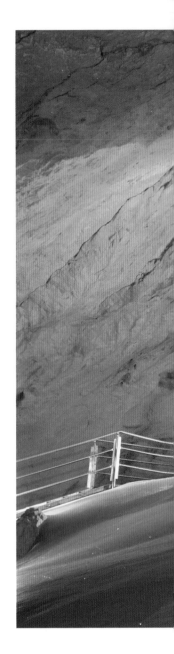

Eisriesenwelt, the Dachstein ice is permanent. Films and trickles of meltwater recycle some of the ice late in the summer when the ground is at its annual warmest, and this is just about matched by each winter's new ice. Huge sheets of layered ice, up to 15 m (50 ft) thick, now form underground glaciers that take up the complete width of some very spacious passages. While newly-formed winter ice decorations in some caves can be very beautiful, and very fragile, the massive scale of the permanent ice deposits in these splendid Austrian caves presents a whole new range of underground processes.

Above: **Underground ice**
A huge ramp of thick ice forms a wide underground glacier, descending an inclined section of the great ancient trunk passage that is the main gallery in the long cave system of the Eisriesenwelt.

Skocjanske Jame █ Slovenia

[GIANT RIVER CAVE]

Opposite: Hanke Bridge

A narrow footbridge carries the

Skocjanske tourist trail across Hanke's

Passage, where the River Reka

tumbles over cascades 45 m (150 ft)

below. Out of sight downstream, the

passage reduces in size so that

floodwaters back up and flow over

the bridge in extreme conditions.

CAVES ABOUND BENEATH the bare limestone hills and endless dolines of the Classical Karst - the Kras of southern Slovenia - but their arterial drain is the underground River Reka. Drainage from adjacent sandstone hills gathers into the river that sinks into the limestone beneath the tiny village of Skocjan, from where it flows underground to the Timavo springs, 35 km (22 miles) away on the Italian coast.

A massive cave passage swallows the River Reka, but almost immediately breaks out into daylight where the river crosses the floors of the Mala and Velika Dolinas - two huge collapses of the cave roof. Each is a rocky sinkhole well over 100 m (300 ft) wide and deep, so they are described as tiankengs (see p.62). The view across them, dwarfing the spire of Skocjan's church, is one of the world's classic karst sights. From the floor of Velika Dolina, the Reka cascades into the giant cave passages of Skocjanske Jame. Exploration of this incredible cave started in 1839 and continued for over 50 years. It involved heroic adventures in almost total darkness, where candles were grossly inadequate as the only lighting. Wooden boats were dragged in to cross the lakes, and traverse cables were bolted high on the passage walls to climb past thundering waterfalls. In later years, pathways were created along rock ledges, and a tunnel was cut in 1933, opening a circular route to create a superb show cave. Today, small groups of visitors gather in the village above, to walk the trail through what still vies (with Carlsbad Caverns, see p.30) for the status of the finest show cave in the world.

The tourist trail enters the tunnel in the floor of Dolina Globocak, another huge sinkhole, but this one is dry and with gently graded slopes. And the tunnel pops out into the end of Tiha Jama - a splendid passage richly decorated with massive stalactites and stalagmites. It is a delight to wander between the calcite towers of Tiha, but this gives no hint of what is to come. Without warning, the dry gallery opens out into the gigantic river passage, not at floor level but on a balcony nearly 80 m (260 ft) above the cave river. This really is making a grand entrance - the view is dramatic and awe-inspiring.

[The big river passage]

From its airy entry, the tourist trail winds round ledges high above the river, and then crosses a bridge, still more than 40 m (130 ft) above the rumbling river which heads off downstream into the tall Hanke's Passage. The river's original route was out along the Tiha cave, along a totally flooded and almost level passage whose continuation lies unseen behind the debris slopes in Dolina Globocak. But then the water found a more steeply descending route towards the sea, along Hanke's Passage. Through much of the last few million years, the cave river cut deeply into its floor, to create the huge canyon in the upstream part of Skocjanske Jame and the narrower canyon of Hanke's Passage. This extends downstream through some large chambers to a flooded section, not yet fully explored. Beyond, its course is accessible only in bits of large passage now known as separate caves, each reached through a deep shaft from the Kras plateau.

From the bridge, the tourist trail heads upstream, first along ledges high on the walls and then almost at river level. The show cave floodlights give only glimpses of this truly massive river passage, but with views both from the high-level path and from the river bank, it is possible to appreciate the sheer size of this river cave. It is well over 50 m (160 ft) high and wide, and is unmatched in any other cave developed with tourist paths and lights. The Reka is a powerful river, and has been flowing this way for an awfully long time; this is the very best of underground erosion. Upstream, the path re-climbs the passage walls and swings into a dry, high-level tunnel where the active canyon has looped away from its ancestral passage. Terraced gour pools (see p.6) cascade down one wall, before the spacious tunnel heads for daylight - emerging in the wall of the Velika Dolina. This huge rock collapse has occurred at a convergence of cave passages each over 20 m (60 ft) high and wide. The tourist trail emerges from one, and a high-level cavern continues into the northern wall until it is choked by banks of ancient silt, where charcoal and artefacts indicate that Neolithic man lived in this spacious rock shelter. Far below, the Reka emerges from one cave, drops over a cascade, and disappears into the main Skocjanske Jame. A narrow path winds skywards, but a cable-hauled railway takes visitors back up to the plateau - the end of the trip through a magnificent cave.

Below: **Giant canyon**
Blocks of eroded limestone protrude from the sweeping expanses of water in the giant canyon passage that is the highlight of Skocjanske Jame.

Krizna Jama | Slovenia

[BEAUTIFUL CAVE LAKES]

WATER IS THE FUNDAMENTAL COMPONENT of the cave environment. Not only does it create the caves in the first place, and then create all the decorations, but water also adds so much to the visual delights of the underground. Roaring waterfalls offer elements of drama and excitement, but the placid waters of still lakes can offer the most magical and delightful of cave experiences. None is finer than the splendid chain of lakes that beckon the visitor through the beautiful galleries and decorated chambers of Krizna Jama.

The inland karst of Slovenia contains a host of splendid caves, where rivers drain through the low limestone hills between poljes - those karst valleys distinguished by their great flat floors. Adjacent to the Planinsko Polje, Postojna is the best known of the caves because it has long been so well developed for tourists. Further east, the Cerknisko Polje forms a lake each winter, but dwindles to a marsh each summer, and one of the streams draining into it comes through the Krizna Jama. A dry entrance in the pine forest gives no hint of what lies inside. Beyond the first few boulder-strewn chambers, the passage shrinks down to a clean-washed, well-rounded tunnel about 8 m (25 ft) in diameter and half full of water. Its deep lake extends into the distance and out of sight round a gentle corner.

This is the first of the 22 lakes that give Krizna Jama its distinctive character. The way on is by boat, and two-man rubber dinghies have replaced the wooden boats that were used by the earlier explorers. Gently paddling your own boat through these half-flooded limestone tunnels is the only way to explore this lovely cave. There is something new around every corner; stalactites, grottoes and flowstone cascades are each enjoyed at the supremely leisurely pace of non-powered boating.

The cave passages of Krizna Jama were largely formed as totally flooded tunnels that carried large flows of water. Then surface lowering (see p.5) left the caves above the water table, when they were partially filled with masses of stalactites and stalagmites. Only a small stream

GREAT CAVES OF THE WORLD | EUROPE

now flows through the cave, and its saturated waters have deposited calcite on the floor to form numerous low gour barriers (see p.6). It is these that now hold back the chain of lakes, and the boats have to be dragged up and over each rounded dam of sparkling white calcite to then re-launch into the next lake; rubber dinghies are so much easier than the heavy wooden boats of yesteryear.

[**Kalvarija**] Thirteen lakes into the cave, a sweeping bend with wall-to-wall water swings round into the chamber of Kalvarija - the classic symbol of Krizna Jama. A great pile of breakdown blocks (see p.7) fallen from the roof creates a ramp leading up from the water's edge and is half-covered with a fabulous array of tall, thin stalagmites. Stalactites hang from the chamber roof, and some have merged with stalagmites from below to create grand, tiered columns. Two streams trickle in, each flowing silently over banks of calcite, forever building the dams that hold back the giant gour pools. Kalvarija offers one of those immensely pleasing underground scenes with the perfect balance between limestone bedrock, calcite stalagmites and the water that created everything in the first place.

Beyond Kalvarija, the far reaches of Krizna Jama have five converging stream passages. The Pisani branch continues upstream with another seven long lakes in an equally spectacular tunnel. It finally breaks out into the largest chamber in the cave, over 150 m (500 ft) long; Kristalna Gora translates as the Crystal Mountain, whose massive pile of breakdown blocks is again decorated with a forest of stalagmites. Not far beyond the great chamber, the accessible passage divides to a slightly ignominious ending where the stream emerges from a choke of stalagmites.

By world standards, Krizna Jama is neither long nor deep, but its long lakes and splendid stalagmite chambers make it a true classic. It is also fitting that this last of the great caves lies in the limestone hills of Slovenia, where the scientific study of caves all began, in what is now known as the Classical Karst. Krizna was first explored back in 1832. It was an exciting discovery back then, and it remains to this day as one of the world's great caves.

Opposite: Kalvarija
The classic view of the Kalvarija chamber that is seen as the visiting caver paddles his boat in, across the cave's thirteenth lake. A long history of cave enlargement, roof collapse and stalagmite growth has created the textbook image of a cave.

Below: Cave boatmen
Long lakes, deep water, an arched limestone roof and two men in a boat are the features of Krizna Jama that make the cave so unusual and so memorable.

[Further information] Drawing a compromise between heavy, erudite, academic texts and humorous ramblings on cavers' adventures,

the following five titles are among the best to offer a grand picture of the world of caves.

Arthur Palmer, 2007. *Cave Geology*. Cave Books, Dayton Ohio, ISBN 0-939748-66-2. 454 pp. The best explanation of how, where and why caves are formed and developed.

Howard Beck, 2003. *Beneath the Cloud Forests*. Speleo Projects: Basel, ISBN 3-908495-11-3. The full exciting story of exploration of the giant caves in Papua New Guinea.

John Gunn, (Ed.), 2004. *Encyclopedia of Caves and Karst Science*. Fitzroy Dearborn. New York and London, ISBN 1-57958-399-7, 902 pp. Accessible information on every aspect of caves.

Michael Taylor, (Ed.), 1991. *Lechuguilla, Jewel of the Underground*. Speleo Projects, Basel. ISBN 0-909158-55-2, 144 pp. The most beautiful photographs of the world's most beautiful cave.

Tony Waltham, 2007. *The Yorkshire Dales, Landscape and Geology*. Crowood, Marlborough, ISBN 978-1-86126-972-0. Caves in the wider context of a major cave region.

National caving organisations websites include www.bcra.org.uk for Britain, www.caves.org for USA, www.caves.org.au for Australia, and www.uis-speleo.org for the international network.

[Visiting the caves]

Of the sites described in these pages, nine are partly developed as show caves with tourist access:

Sterkfontein Cave: *show cave, www.sterkfontein-caves.co.za*

Mammoth Cave: *show cave, www.nps.gov/maca*

Carlsbad Caverns: *show cave, www.nps.gov/cave*

Akiyoshi-do: *show cave, www.apike.ca/japan_akiyoshido*

Perak Tong Cave: *open access to active temple cave, www.perak.info/kinta/Perak_Tong_Chinese_Temple*

Caves of Mulu: *Deer and Clearwater Caves have tourist access, www.forestry.sarawak.gov.my/forweb/np/np/mulu*

Waitomo Caves: *show caves at Waitomo, www.waitomo.com/waitomo-glowworm-caves.aspx and Mangawhitikau, www.glowworm.co.nz*

Ice Caves of Austria: *show caves at Eisriesenwelt, www.eisriesenwelt.at and Dachstein Reiseneishöhle, www.dachsteinwelterbe.at*

Skocjanske Jame: *show cave, www.park-skocjanske-jame.si/eng*

Another six of the caves have parts that can be visited on adventure trips with local guides:

Sof Omar Cave: *guides live in the village beside the river sink entrance.*

Gruta do Janelão: *access to Peruaçu National Park may be arranged by local tour companies.*

Difeng Dong: *there is footpath access to the floor of Xiaozhai Tiankeng.*

Tham Hinboun: *tour companies in Laos can offer tours that include a trip through the cave.*

Gaping Gill: *winch descents booked through www.bpc-cave.org.uk and www.cravenpotholeclub.org*

Krizna Jama: *visits arranged through www.krizna-jama.si*

The remaining caves are only accessible to experienced and fully equipped cavers.

[Acknowledgements] Thanks to Art Palmer for checking the text, and to friends who have made visits to the caves possible.

[Picture credits] Front cover © Geophotos; back cover photo: Borut Lozej, © Park Skocjan Caves; **pp.4, 6, 7, 10, 15, 16, 17, 18-19, 23, 26, 31, 38, 39, 45, 46, 50, 51, 53, 54-55, 63, 68-69, 70, 73, 74, 76, 83, 84-5, 90, 91, 94, 102, 111** © Geophotos; **p.5** © Robbie Shone Photography; **p.9** © John Gunn; **p.12** © The Natural History Museum; **p.13** © Jae Jong Kwak; **p.20** © Art Palmer; **p.22** © Dave Bunnell; **p.25** © Art Palmer; **p.28** © Mark Tracy; **p.29** © Art Palmer; **pp.32, 33, 34** © Peter Jones; **p.35** © Art Palmer; **p.36** © Rob Ratkowski/Maui; **p.41** © Art Palmer; **p.42** © Martyn Farr/Farrworld; **p.44** Photo: Brian Zane/Jamaican Caves Organisation. © Brian Zane; **p.49** © Andy Eavis; **pp.56, 57** © Call of the Abyss Project; **pp. 58, 60** © Project NAMAK, photo: Marek Audy and Richard Bouda; **p.64** © Tony Baker; **p.67** Courtesy Akiyoshi-do Museum, © Geophotos; **p.75** © Jerry Wooldridge; **pp.78-9** © Robbie Shone Photography; **pp.80, 81** © Dave Gill; **p.87** © Spellbound Tours, Waitomo Caves; **p.89** © Paul Deakin; **pp.93, 95** © French Ministry of Culture and Communication, Cultural Affairs - Rhône-Alpes; **pp.97, 98** © Paul Deakin; **p.99** © Tony Baker; **pp.101, 103** © Eisriesenwelt; **pp.105, 106, 107** Photo: Borut Lozej, © Park Skocjan Caves; **p.109** © Guy Edwardes; **p.110** © Jerry Wooldridge.

Every effort has been made to contact and accurately credit all copyright holders. If we have been unsuccessful, we apologise and welcome correction for future editions and reprints.

Front cover: Gruta do Janelão, Brazil; *back cover:* Skocjanske Jame, Slovenia.